Workbook for
Structure and *Function*
of the *Human Body*

Workbook for
Structure and Function of the Human Body

Ruth Lundeen Memmler, M.D.
Professor Emeritus, Life Sciences;
formerly Coordinator, Health, Life Sciences and Nursing,
East Los Angeles College, Los Angeles

and

Dena Lin Wood, R.N., B.S., P.H.N.
Assistant Head Nurse, Los Angeles County—
University of Southern California
Medical Center, Los Angeles

Illustrated by Anthony Ravielli

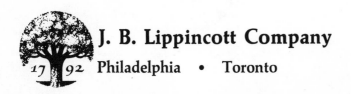
J. B. Lippincott Company
Philadelphia • Toronto

ISBN 0-397-54198-8

7 9 8

Preface

This workbook is designed to aid the beginning student in the study of fundamental principles involving the human organism as they are presented in the text *Structure and Function of the Human Body*, second edition. It may also be useful in connection with other textbooks that include the anatomical and physiological principles related to the human body.

Pronouncing the scientific words while writing them in answer to the questions in the workbook will help improve the student's understanding of the material. The student will learn most by proceeding in an independent fashion with no copying of the work of classmates.

The questions are written with the aim of assisting the student in the learning process. They are not intended as a test of his knowledge. Sometimes the answer to one question is purposely included in another interrogation in order to reinforce the ideas that are presented. The student should analyze the facts and try to place them in a meaningful order.

The authors are grateful for suggestions and help given by colleagues and by the staff of the J. B. Lippincott Company, especially Bernice Heller and David T. Miller.

Contents

ONE
The General Plan of the Human Body   1

TWO
Cells, Tissues and Membranes   11

THREE
The Blood   27

FOUR
Body Temperature and Its Regulation   37

FIVE
The Skin   43

SIX
Bones, Joints and Muscles   49

SEVEN
The Brain, the Spinal Cord and the Nerves   69

EIGHT
The Sensory System   91

NINE
The Heart   103

TEN
Blood Vessels and Blood Circulation   109

ELEVEN
The Lymphatic System and Lymphoid Tissue   129

TWELVE
The Digestive System   137

THIRTEEN
The Respiratory System   151

FOURTEEN
The Urinary System   159

FIFTEEN
Glands and Hormones   171

SIXTEEN
The Reproductive System   179

CHAPTER ONE

The General Plan of the Human Body

I. OVERVIEW

In order to understand the general organization of the human body we may begin with the smallest "bricks," or *cells*, which are the basic structural units of all living things, and which are composed of protoplasm. Although there are numerous types of cells, which differ in size and attributes according to function, all have a common basic structure. The outer *cell membrane* contains the main substance, or *cytoplasm*; in the center of the cell is the *nucleus*, which contains the still smaller *nucleolus*. Cells combine to form tissues that in turn form *organs*; these organs form *systems*.

It is essential that a special set of terms be learned in order to locate parts and to relate the various parts to each other. Imaginary lines called *planes of division* separate parts of the body into *regions* in much the same way that the equator, the Tropics of Cancer and Capricorn and the Arctic and Antarctic Circles divide the earth into zones. Further divisions of the earth by lines of latitude and longitude make it possible to pinpoint locations accurately. Similarly, separation into areas and regions within the body, together with the use of the special terminology for directions and locations, makes it possible to describe an area within the human body with considerable accuracy.

The logical decimal type metric system is now replacing all other systems of measurement. Students must learn to think metric.

II. TOPICS FOR REVIEW

1. microscopic structure
   a. cells and protoplasm
   b. nucleus and other cell parts
2. body planes, cavities and systems
3. body directions
   a. anatomical position
   b. superior and inferior
   c. ventral and dorsal
   d. anterior and posterior
   e. cranial and caudal
   f. medial and lateral
   g. distal and proximal

III. MATCHING EXERCISES

Matching only within each group, print the answer in the space provided.

Group A

organs	tissues	cells
systems	cytoplasm	nucleus
protoplasm		

1. The substance of which all living things are composed is _protoplasm_.

2. The central, usually oval, part of a cell is the............ _nucleus_.

3. A combination of specialized groups of cells forms....... _tissues_.

4. The main substance of the cell outside the nucleus is the.. _cytoplasm_.

5. A combination of various tissues form parts having a special function called .... _organs_.

6. Several different parts and organs grouped together for specific functions form .... _systems_.

7. The building blocks of which living organisms are made are called .... _cells_.

Group B

epigastrium	ventral	distal
umbilicus	lateral	medial
proximal	thoracic region	transverse

1. To indicate nearness to the midsagittal plane use the word _medial_.

2. A part that is away from the midline (or toward the side) is .... _lateral_.

3. To indicate that a part is near or toward the point of origin use .... _proximal_.

4. A part that is away from the point of origin is........... _distal_.

5. A horizontal or cross section is also said to be.......... _transverse_.

6. The central region of the abdomen just below the breast bone is the .... _epigastrium_.

7. Another name for the navel is the...................... _umbilicus_.

8. The upper or chest portion of the ventral body cavities is the .... _thoracic region_.

9. The word that means toward the belly surface is........ _ventral_.

2

Group C

caudal	cranial	posterior
ventral	proximal	lateral

1. To say, toward the origin of a part, use the word........ *proximal*.

2. To indicate that a part is toward the rear, use........... *posterior*.

3. The word that means nearer the tail region is.......... *Caudal*.

4. To indicate that a part is nearer the head use the word... *cranial*.

5. To show that a part is toward the side use the word..... *lateral*.

6. To show that a part is nearer the belly area use........ *ventral*.

Group D

urinary	integumentary	skeletal
endocrine	reproductive	respiratory

1. The system that includes the hair, nails and skin is the... *integumentary*.

2. The bones, joints and related parts form the system called the ..... *skeletal*.

3. Another name for the excretory system is the........... *urinary*.

4. The system of scattered organs that produce hormones is called the ..... *endocrine*.

5. The system that includes the sex organs is the.......... *reproductive*.

6. The lungs and bronchial tubes form the system called the *respitory*.

Group E

spinal canal	~~cranial cavity~~	diaphragm
~~midsagittal~~	chromatin network	metric system

1. A logical decimal type of measurements is the.......... *metric system*.

2. The plane that divides the body into right and left halves is the ..... *midsagittal*.

3. The lower elongated part of the dorsal body cavity is the *spinal canal*.

4. The muscular partition between the 2 ventral body cavities is the ..... *diaphram*.

5. Deeply staining granules within the cell nucleus form the *Chromatin network*.

6. The upper part of the dorsal body cavity is the.......... *Cranial cavity*.

3

IV. LABELING

For each of the following illustrations, print the name or names of each labeled part on the numbered lines.

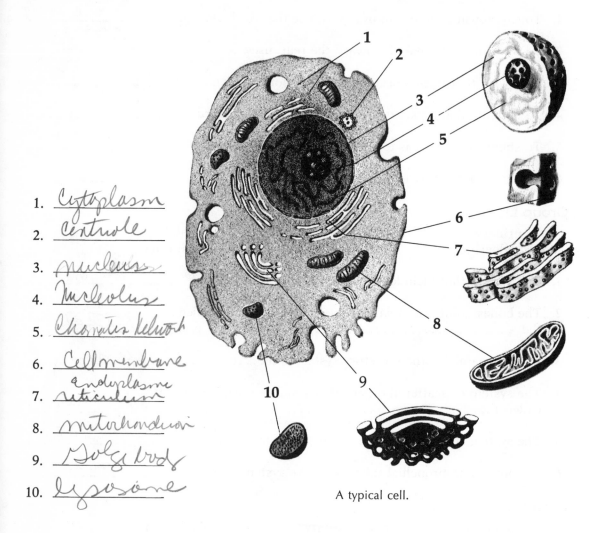

A typical cell.

1. Cytoplasm
2. Centriole
3. nucleus
4. nucleolus
5. Chromatin Network
6. Cell membrane
7. endoplasmic reticulum
8. mitochondria
9. Golgi body
10. lysosome

1. Cranial Superior
2. transverse plane
3. midsagittal plane
4. inferior
5. dorsal posterior
6. ventral anterior
7. frontal plane

4

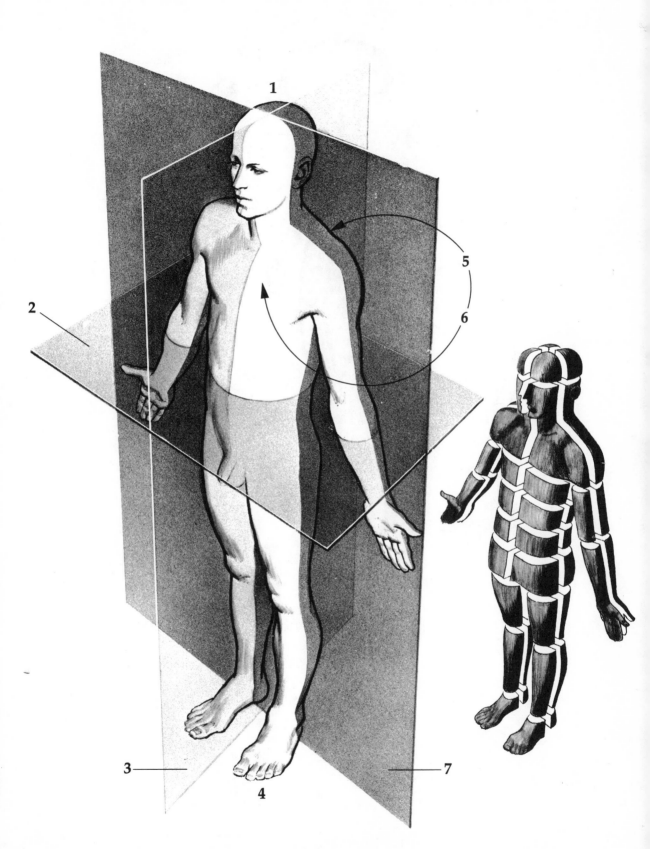

Body planes and directions.

5

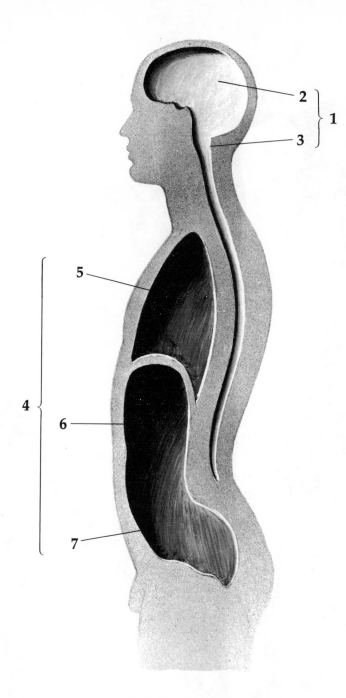

Side view of body cavities.

1. _dorsal body cavities_   5. _____

2. _Cranial_   6. _____

3. _____   7. _____

4. _____

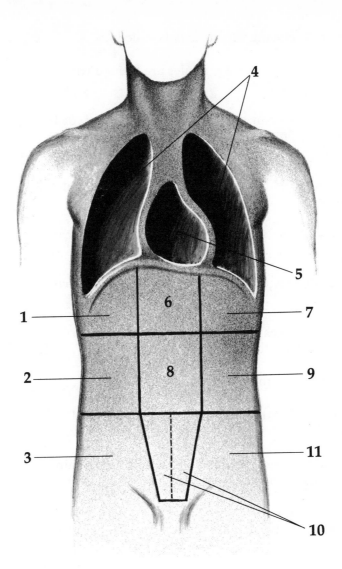

Front view of body cavities and the regions of the abdomen.

1. _____   7. _____

2. _____   8. _____

3. _____   9. _____

4. _____   10. _____

5. _____   11. _____

6. _____

V. COMPLETION EXERCISE

Print the word or phrase that correctly completes the sentence.

1. The protoplasm enclosing the nucleus is called......... _____.

2. The deeply staining chromatin granules are located within the part of a cell called the........................ _____.

3. Regions and directions in the body are described according to the position in which the body is upright with the palms facing forward. This is called the.............. _____.

4. The midline plane that divides the body into right and left halves is the.................................. _____.

5. Planes that divide the body into upper and lower parts are called .... _____.

6. The plane that divides the body into front and rear parts is the .... _____.

7. The space that encloses the brain and spinal cord forms one continuous cavity, the........................... _____.

8. The space that houses the brain is the................ _____.

9. The elongated canal that contains the spinal cord is known as the .... _____.

10. The ventral body cavities include an upper space containing the lungs, heart, and the large blood vessels, which is called the .... _____.

11. The lower ventral body cavity is quite large and is called the .... _____.

12. The large ventral body cavities are separated from each other by a muscular partition, the.................... _____.

13. The large lower ventral body cavity is subdivided into nine regions including 3 nearer the midline. The uppermost of these midline areas is the.................... _____.

14. The standard metric measurement for volume, slightly greater than a quart, is called a..................... _____.

VI. PRACTICAL APPLICATIONS

Study each discussion. Then print the appropriate word or phrase in the space provided.

Group A

1. The gallbladder is located just below the liver. The directional terms that best describe this relationship include.. _____.

8

2. The kidneys are located behind the other abdominal organs. This relationship may be described as.......... ———————.

3. The tips of the fingers and toes are farthest from the region of origin of these digits so they are said to be the most .... ———————.

4. The entrance area of the stomach is nearest the point of origin or beginning of the stomach so this part is said to be .... ———————.

5. The ears are located away from the midsagittal plane or toward the side so they are described as being.......... ———————.

6. The head of the pancreas is nearer the midsagittal plane than its tail portion, so the head part is more.......... ———————.

7. The diaphragm is above the abdominal organs; it may be described as .... ———————.

Group B

On the ward in which postoperative patients were being cared for you were asked to study certain cases and answer the following questions.

1. Mr. A had an incision for an appendectomy. The area of the abdomen in which the appendix is located is in the lower right side and is known as the.................. ———————.

2. Mrs. B had a history of gallstones. The operation to remove these stones involved the upper right part of the abdominal cavity, or the............................. ———————.

3. Miss C was injured in an automobile accident. In addition to a number of fractures she suffered a ruptured urinary bladder. The area in the lower midline part of the abdomen is the .... ———————.

4. Mr. B required an extensive exploratory operation that involved an incision through the navel. This portion of the abdomen is the................................. ———————.

CHAPTER TWO

Cells, Tissues and Membranes

I. OVERVIEW

While the organization of matter into atoms and molecules involves submicroscopic structures, the composition of protoplasm includes microscopic and larger parts beginning with *cells* and their *organelles*. Whether a cell exists alone or as one unit of a structure, its work goes on ceaselessly through its *organelles* ("little organs"). Among these are the *mitochondria*, which contain the catalytic *enzymes*. Through *mitosis*, new cells are constantly being formed. They receive nourishment, generate heat and energy, get rid of waste products and discharge secretions by means of the *semipermeable* cell wall; this entire process is known as *metabolism*.

The tissues are composed of specialized groups of cells having a common purpose, and are classified into 4 main groups: *epithelium, connective tissue, nerve tissue* and *muscle tissue*. The simplest combination of tissues is the *membrane*—the thin sheet of material that separates 2 groups of substances. In the group of *epithelial* membranes are 2 subgroups: the *mucous* membranes lining tubes and other spaces that open to the outside of the body, and the *serous* membranes that line closed cavities inside the body. The second group of membranes, *connective tissue* membranes, is also divided into 2 subgroups: the *fascial* membranes anchor and support the organs, and the *skeletal* membranes cover bone and cartilage.

II. TOPICS FOR REVIEW

1. cell organelles including mitochondria
2. DNA, RNA and enzymes
3. cell division or mitosis
4. the physical processes of diffusion, osmosis and filtration
5. characteristics of tissues in general
6. the 4 main kinds of tissues and their anatomical characteristics
7. functions of each of the 4 types of tissue
8. definitions: types of membranes
9. main categories of membranes
    a. epithelial membrane
    b. connective tissue membrane

10. subgroups of epithelial membranes
    a. characteristics and function of mucous membranes; examples
    b. characteristics and function of serous membranes; examples
11. subgroups of connective tissue membranes
    a. characteristics and function of fascial membranes; examples
    b. characteristics and function of skeletal membranes; examples

III. MATCHING EXERCISES

Matching only within each group, print the answer in the space provided.

Group A

energy	glucose	amino acids
osmosis	diffusion	filtration
chromosomes	organelles	enzymes

1. Certain complex proteins that act as catalytic agents are
   classified as .... _____.

2. The microscopic structures that are present in practically
   all living cells and that regulate a variety of functions
   within the cells are the................................. _____.

3. Molecules move from an area of relatively high concentration to an area of lower concentration in the process of _____.

4. Rod-shaped structures that are deeply staining and are
   found only during cell division are known as............ _____.

5. The passage of a solvent through a semipermeable membrane from an area of lower concentration to one that is
   higher is the process of................................. _____.

6. The passage of water with its dissolved substances through
   a membrane as a result of a greater mechanical force on
   one side is the process called........................... _____.

7. During digestion carbohydrates are changed to a simple
   sugar called .... _____.

8. The end products of protein digestion are.............. _____.

9. The metabolism of foods may produce the capacity for
   action, namely .... _____.

Group B

resting stage	adipose	secretions
tendons	cilia	mitochondria
cartilage	bone	inheritance

1. The chromatin material of the cell nucleus exists as granules during the cell's usual.......................... —————————.

2. The genes, which are contained in the chromosomes, regulate .... —————————.

3. Enzymes stimulate chemical activities within the cell. They occur inside the organelles called................ —————————.

4. An important function of epithelium is the production of ————————.

5. The tiny hairlike protoplasmic extensions that project from epithelium which help to prevent lung damage by keeping the airways clear are the...................... —————————.

6. The storing up of fat, heat insulation and padding of various structures are functions of the type of connective tissue called .... —————————.

7. The strong, cablelike connective tissue cords that connect muscles to bones are..................................... —————————.

8. One of the hard connective tissues that has the important function of acting as a shock absorber and as a bearing surface to reduce friction between moving parts is.... —————————.

9. Osseous tissue is similar to cartilage in its cellular structure. Cartilage may gradually become impregnated with calcium salts to form .... —————————.

Group C

catalysts	tissue fluid	edema
neurons	semipermeable	dehydration
water	tissues	metabolism
physical and chemical changes	DNA	RNA

1. Substances that increase the speed of chemical reactions without being changed themselves are................... —————————.

2. Within the cell, the production of heat and energy, new protoplasm and waste products is known as............ —————————.

3. Organized groups of cells that are of the same type and that have a common purpose form...................... —————————.

4. The basic structural units of nerve tissue are the........ —————————.

5. From 60 percent to 99 percent of body tissues are made up of the abundant compound......................... —————————.

13

6. Bathing the tissues is a slightly salty solution, the....... _____.

7. In conditions of excessive fluid loss the tissues suffer from _____.

8. The chief component of chromosomes is a complex molecule called .... _____.

9. Instructions given by the DNA molecule are carried from the nucleus to other parts of the cell and to all parts of the body by another complex molecule called......... _____.

10. A puffiness of tissues due to an abnormal accumulation of fluid is found in the condition called................... _____.

11. Metabolism is the term that describes a combination of activities within the cell including all the.............. _____.

12. The cell wall is selective, that is, it permits some substances to enter the cell but prevents passage of others. Therefore, it is said to be............................ _____.

Group D

myocardium	fibers	connective tissue
cell division	neurilemma	myelin
voluntary muscle	spasm	visceral muscle

1. Areolar, adipose and osseous tissue all act as the body's supporting fabric and are therefore classified as......... _____.

2. The basic structural unit of nerve tissue, the neuron, consists of a nerve cell body plus small branches, which are called .... _____.

3. The ability of certain nerves to repair themselves is due to the presence of...................................... _____.

4. Like telephone wires, nerve fibers are encased in a protective covering, or sheath. This fatty insulating material is called .... _____.

5. The thickest layer of the heart wall is formed by cardiac muscle or .... _____.

6. Certain diseases are characterized by abnormal muscle contractions. A single sudden violent contraction is classified as a .... _____.

7. Muscle tissue is classified into 3 types. That which forms the walls of the organs within the ventral body cavities is called .... _____.

8. Skeletal muscle provides for the movement of the body. It is therefore described as............................ ——————————.

9. Even after maximum growth is attained there is still a continual (though slower) process of cell production by means of .... ——————————.

Group E

omentum	connective tissue	mesentery
liver	epithelium	urinary bladder
pancreas	cul de sac	visceral pleura

1. The main tissue of many protective coverings and of the linings of the respiratory and digestive tracts, among others, is .... ——————————.

2. Repair of damaged nerve and muscle tissue is accomplished by the growth of............................ ——————————.

3. The uppermost and the largest of the organs inside the peritoneal cavity is the............................ ——————————.

4. Extending downward from the stomach is an apronlike fold of peritoneum called the........................ ——————————.

5. In the midline behind the stomach in an area called the retroperitoneal space are the duodenum and the glandular ——————————.

6. Connecting the small intestine to the back wall is a double-layered peritoneal structure, the................ ——————————.

7. In the pelvis below the peritoneal cavity is a hollow organ, the .... ——————————.

8. The serous membrane that covers the lung surfaces is called the .... ——————————.

9. The extension of the greater peritoneal cavity downward behind the reproductive organs is called the............. ——————————.

Group F

pleurae	cell wall	serous membranes
pericardium	membrane	mucous membranes
peritoneum	lubricants	fascial membranes

1. Any thin sheet of material that separates 2 or more groups of substances is classified as a.................. ——————————.

2. The membrane that permits certain substances to enter the cell and certain substances to pass out of it is the.... ——————————.

3. The membranes that line the so-called closed cavities within the body are.................................. _____.

4. The tough membranes composed entirely of connective tissue which serve to anchor and support organs are the _____.

5. The linings of tubes and spaces that are connected with the outside are largely epithelial. They are.............. _____.

6. The membranes that form the 2 separate sacs for the lungs are known as the............................... _____.

7. The special sac that encloses the heart is known as the... _____.

8. The serous membrane of the abdominal cavity is the largest of its kind and it is called the................... _____.

9. An important function of most epithelial membranes is to produce fluids that serve as......................... _____.

Group G

superficial fascia	periosteum	synovial membranes
parietal layer	perichondrium	mucous membranes
mesothelium	capsules	

1. Membranous connective tissue envelopes that enclose organs are called .... _____.

2. The tough connective tissue membrane that serves as bone covering is the................................ _____.

3. Covering cartilage is a membrane similar to that covering bone. It is called.................................. _____.

4. Secretions produced by the linings of joint cavities act as lubricants to reduce friction between the ends of bones. These linings are .... _____.

5. The linings of the various parts of the respiratory tract are all .... _____.

6. The tissue that underlies the skin is known as the....... _____.

7. The part of a serous membrane that is attached to the wall of a cavity or sac is the.......................... _____.

8. Movements of organs occur with a minimum of friction because of the presence of a type of epithelium called.... _____.

IV. LABELING

For each of the following illustrations, print the name or names of each labeled part on the numbered lines.

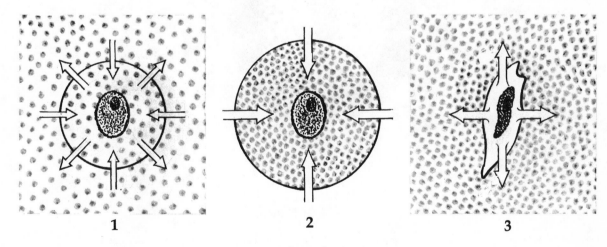

1   2   3

Osmosis.

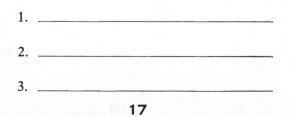

1. _____

2. _____

3. _____

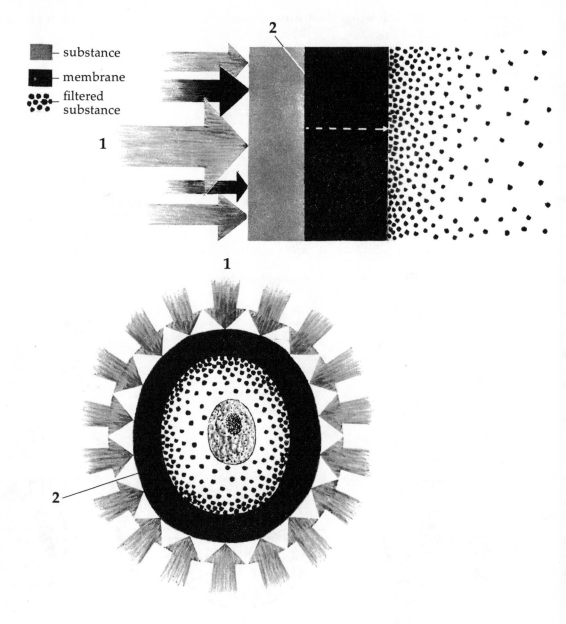

substance
membrane
filtered substance

2

1

1

2

Filtration.

1. _____

2. _____

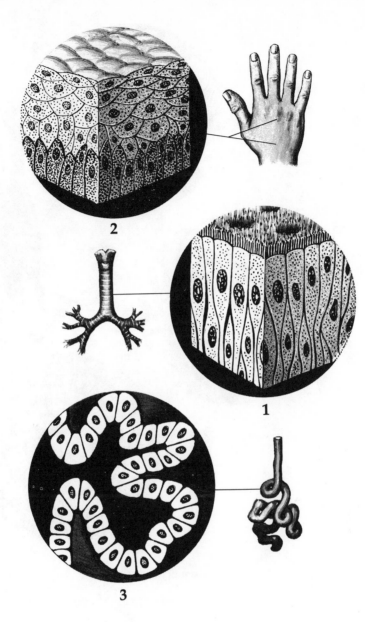

Three types of epithelium.

1. _____

2. _____

3. _____

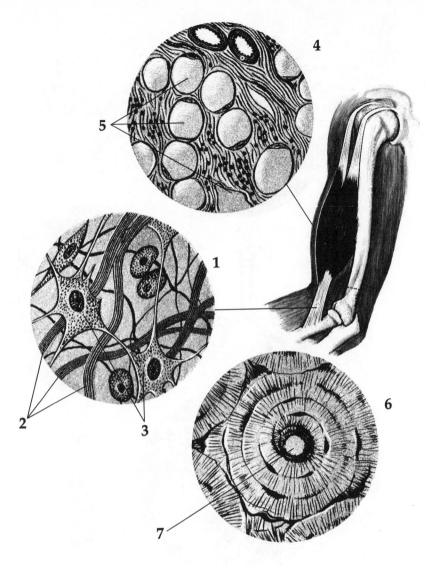

Connective tissue.

1. _____

2. _____

3. _____

4. _____

5. _____

6. _____

7. _____

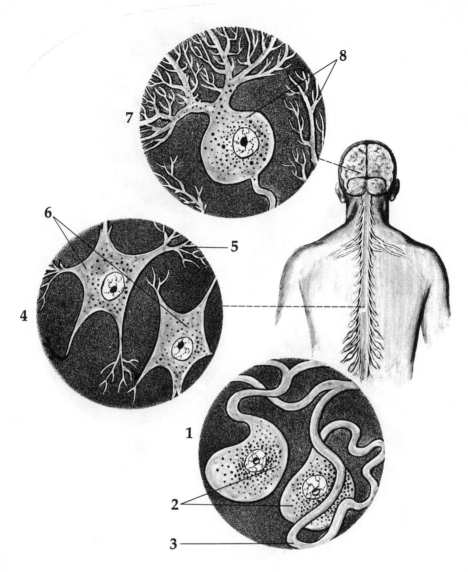

Nerve tissue.

1. _____

2. _____

3. _____

4. _____

5. _____

6. _____

7. _____

8. _____

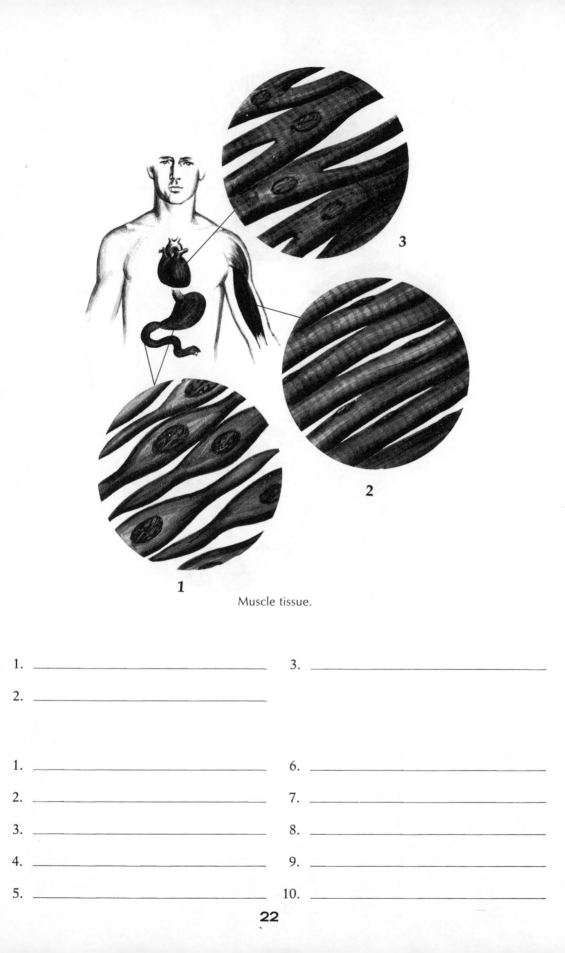

Muscle tissue.

1. _____

2. _____

3. _____

1. _____

2. _____

3. _____

4. _____

5. _____

6. _____

7. _____

8. _____

9. _____

10. _____

11. _____   14. _____

12. _____   15. _____

13. _____   16. _____

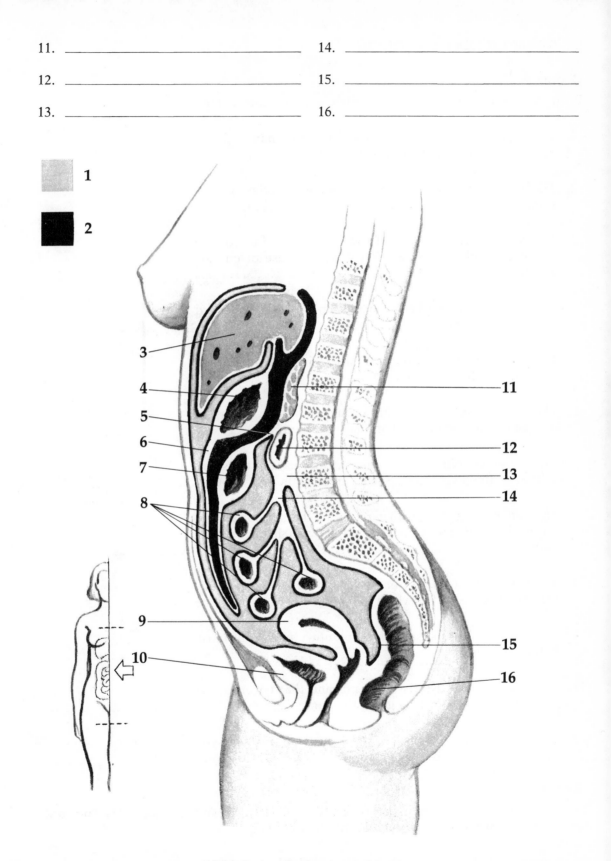

Abdominal cavity showing peritoneum.

23

V. COMPLETION EXERCISE

Group A

Print the word or phrase that correctly completes the sentence.

1. The fundamental building blocks of the human body, as in other living organisms, are the.................... _____.

2. The hereditary units or factors, located within the chromosomes of the nucleus, are called...................... _____.

3. The rod-shaped organelles of the cell, responsible for the chemical combinations leading to the release of energy, are the .... _____.

4. Chemical reactions that go on within a cell are speeded up by catalytic agents called......................... _____.

5. The special connective tissue characteristic of the central nervous system is called.............................. _____.

6. The indirect cell division that occurs in human beings as well as in many plants and animals is called............ _____.

7. Normal saline solution has a concentration about the same as that of the body fluids. Such a solution is said to be... _____.

8. The broad general term that refers to the sum of all the physical and chemical processes that occur in the human body is .... _____.

9. The basic structural unit of nerve tissue is the nerve cell, the scientific name for which is..................... _____.

10. Another term for the smooth involuntary muscle of most hollow organs is .... _____.

11. The tough connective tissue membrane that covers most most parts of all bones is given the name.............. _____.

12. A lubricant that reduces friction between the ends of bones is produced by the............................ _____.

Group B

During digestion food substances are changed into simpler smaller molecules. Indicate specifically what these are by completing the following:

1. Sugars and starches are changed into soluble........... _____.

24

2. Fats are changed into fatty acids and . . . .   —————————————.

3. Proteins are reduced to building blocks called . . . .   —————————————.

VI. PRACTICAL APPLICATIONS

Study each discussion. Then print the appropriate word or phrase in the space provided.

Group A

Observations you might make while touring a hospital laboratory include the following.

1. The janitor in the laboratory was using a cleaning solution that contained ammonia. You will recall that this would cause ammonia molecules to spread throughout the room. This movement of molecules from an area of high concentration to other areas where concentration is low is called . . . .   —————————————.

2. One of the laboratory technicians was trying to separate solid particles from a liquid mixture. He poured the mixture into a paper-lined funnel. The liquid flowed through the funnel while the solids remained behind on the paper. This process is called . . . .   —————————————.

3. A laboratory worker was carefully measuring certain mineral salts in order to prepare a normal saline solution. Normal saline is used to replace lost body fluids because the concentration is nearly the same as that inside the cells. Such a solution is said to be . . . .   —————————————.

4. While doing a complete blood count a technician noted that some of the red blood cells had ruptured. The solutions used were tested to determine whether they were too dilute. Osmosis of water into a cell could be the cause of cell breakage. A too dilute solution is said to be . . . .   —————————————.

5. A student was learning how to do blood smears. Upon examination of the blood with the microscope he found that many red blood cells appeared shrunken. The explanation was that he was proceeding so slowly that the liquid part of the blood was evaporating, leaving a highly concentrated solution. Such a solution is described as being . . . .   —————————————.

Group B

While observing in an outpatient clinic, a student noted the following cases.

1. Baby J experienced difficulty in breathing and copious discharge from his nose. A diagnosis of U.R.I. (upper respiratory infection) was made. The location of the membrane and the type of discharge indicated that the involved membrane was one of the . . . .   —————————————.

2. Mrs. K complained of a swelling in the left groin. She had suffered previously from an infection of bone in the middle back; now it appeared that the infection had traveled along the fibrous covering of some of the back muscles. Such muscle coverings are called.............. _____.

3. Mr. B was concerned about swelling and tenderness over his neck and upper back. His work involved the demolition of old buildings; he had become careless about personal cleanliness. Infection now involved the skin and connective tissue under it. The "sheet" that underlies the skin is called .... _____.

4. Miss G complained of sharp pains in the chest and one side. Her disorder was diagnosed as inflammation of the membrane that forms a sac around each lung. This membrane is the .... _____.

5. Mrs. J had suffered a painful bump on her ankle. The swelling involved the superficial tissues and the fibrous covering of the bone, or the........................... _____.

Group C

While working in an intensive care unit the nurse reported on the following cases.

1. Mr. M had a history of repeated bacterial infections that involved the lining of organs of the reproductive system. Now, because of neglect, his urinary system was also affected. The continuous lining found in the reproductive and urinary systems is classified as.................... _____.

2. Miss G experienced abdominal pains following long-standing infection of the pelvic organs. Connective tissue bands (adhesions) were found to extend throughout the peritoneal surface. The layer of peritoneum that is attached to the organs is called the........................... _____.

3. Miss G complained also of pain with certain motions. This was probably due to the pull of the adhesions on the nerve endings in the abdominal wall. The layer of serosa lining the wall is the.................................. _____.

4. Student N suffered a mild concussion while playing football and it was feared that there might be damage to the brain coverings. These brain and spinal cord coverings are known as .... _____.

CHAPTER THREE

The Blood

I. OVERVIEW

Blood is an important *indicator* of a person's *health*; some understanding of its constituents and their *functions* is needed by all those engaged in health occupations. Blood has the 2 functions of *transporting*, by bringing needed substances such as food materials and oxygen to all the body tissues, and carrying off waste products, and *combating infection*, by defending the body against harmful organisms and maintaining its disease immunity.

The *plasma* of the blood consists of water, protein, carbohydrates, lipids, mineral salts and some other substances which are needed for normal body function. The *formed elements* of the blood, called the *corpuscles*, are composed of the *erythrocytes* (the red cells, which carry oxygen to the tissues by means of their *hemoglobin*), the *leukocytes* (the white cells, which defend the body by engulfing harmful pathogens) and the *platelets* (the thrombocytes, which play an essential role in blood clotting). The corpuscles are mainly formed in the red bone marrow.

One reason for studying the characteristics of blood is to be able to recognize abnormal findings, because such blood findings indicate that certain diseases may be present. Anemias, neoplastic diseases of blood and hemorrhagic disorders are all associated with blood abnormalities.

A second reason is concerned with blood transfusions, which often become necessary in the treatment of various diseases. A procedure known as crossmatching is done to assure that the blood of the patient and that of the donor are compatible.

By adding sodium citrate to blood to prevent clotting, blood can be stored for a number of days to be used in times of emergency. Here, too, the blood must be typed and crossmatched before being transfused.

The presence or absence of the Rh factor, a red cell protein, is also determined by testing—another reason for our need to understand the characteristics of blood. If blood containing the Rh factor (Rh positive) is given to a person whose blood lacks that factor (Rh negative), the recipient may become *sensitized* to the protein; his blood will produce *antibodies* to counteract the foreign substance.

Blood group studies may also provide useful information about paternity in some cases.

In order to make the kinds of determinations mentioned, numerous *blood studies* have been devised. The larger modern laboratories use machines that can count blood cells and others that determine quantities of enzymes, electrolytes and other substances in the blood serum. These machines are accurate and arrive at the answers quickly. The blood smear is used to ascertain the presence of parasites or abnormal red or white blood cells, which is significant in some instances.

II. TOPICS FOR REVIEW

1. purposes of blood
2. blood plasma and its functions
   a. proteins
   b. carbohydrates
   c. lipids
   d. mineral salts
3. the formed elements and their functions
   a. erythrocytes
      (1) structure and function
      (2) purpose of hemoglobin
   b. leukocytes
      structure and function
   c. platelets (thrombocytes)
      origin and function
4. formation of corpuscles
5. blood typing and blood transfusion
   a. blood groups
   b. Rh factor
   c. determining fatherhood
   d. use of blood bank
   e. conditions requiring transfusion
6. blood derivatives
7. disorders of blood
   a. main groups
   b. characteristics
8. blood studies
   a. blood counting machines
   b. hemocytometer
   c. blood slide
   d. blood chemistry machines
   e. clotting and bleeding times

III. MATCHING EXERCISES

Matching only within each group, print the answer in the space provided.

Group A

red marrow	thrombocytes	carbon dioxide
oxygen	plasma	erythrocytes
hemoglobin	leukocytes	protein

1. The liquid part of the blood is known as............... ————————.

2. The red blood cells are called........................ ————————.

3. There are several types of white blood cells or......... ————————.

4. Elements that have to do with clotting include platelets,
   or .... ————————.

5. An important gas that is transported by the blood from the lungs to all parts of the body is.................... _____.

6. The blood carries a waste product to the lungs and this gas is known as........................... _____.

7. After water, the next largest percentage of material of which the plasma is composed is .... _____.

8. An important ingredient of red blood cells is a compound called .... _____.

9. A connective tissue present in bone is the site of formation of most blood cells. The name of this tissue is...... _____.

Group B

serum	fibrinogen	hemolysis
type O	hemoglobin	pathogens
megakaryocytes	type AB	agglutination

1. Oxygen, needed by all the tissues, is transported by the blood constituent .... _____.

2. The platelets are fragments of large cells known as....... _____.

3. As platelets disintegrate they release a chemical that reacts with a plasma protein.......................... _____.

4. The process whereby cells become clumped is known as.. _____.

5. The watery fluid that remains after a clot is removed is known as .... _____.

6. In blood transfusion a dangerous condition that occurs when donor cells are dissolved or go into solution is..... _____.

7. Blood that is not clumped by either anti-A or anti-B serum belongs to the group called.................... _____.

8. If the cells are clumped by both the anti-A and anti-B serums the blood belongs to.......................... _____.

9. The appearance of pus at a body site indicates that the leukocytes are actively involved in the destruction of.... _____.

Group C

mineral salts	fibrin	glucose
no nucleus	lymphoid tissues	ameboid
amino acids	thromboplastin	red marrow
centrifuge	transfusion	hemorrhage

1. A substance that prevents blood coagulation is heparin, while one that triggers the clotting mechanism is........ _____.

2. Circulating in the blood are the protein building blocks absorbed from food. They are called.................... _____.

3. Profuse bleeding is usually referred to as.............. _____.

4. Within the spongy (cancellous) bone of the ends of long bones and within the body of all other bones is the important blood-forming organ called the.................... _____.

5. The transfer of whole blood from one person to another is called .... _____.

6. The solid material formed during clotting by the union of platelets with a chemical from the liver is called......... _____.

7. Separation of blood plasma from the formed elements of blood is accomplished by the use of the................ _____.

8. Leukocytes move out of the blood vessels to an area of infection by a motion described as.................... _____.

9. Normally the blood contains about one part per thousand of a carbohydrate that is a simple sugar called.......... _____.

Group D

smear	leukocytosis	hemoglobinometer
leukopenia	hemocytometer	4.5 to 5.5 million
5,000 to 9,000	hematocrit	differential

1. An apparatus made of several parts and used for counting blood cells is called a............................ _____.

2. Normally, the number of red blood cells per cubic millimeter is .... _____.

3. Normally, the number of white blood cells per cubic millimeter is .... _____.

4. The volume percentage of red blood cells in whole blood is called the .... _____.

5. In most infections, as well as in various other types of illness, the white count may be excessive. This finding is referred to as .... _____.

6. An abnormal reduction of the white blood count to below 5,000 is called .... _____.

7. The apparatus that determines the amount of hemoglobin in the blood is known as the hemometer or............. _____.

8. An estimation of the percentage of each type of white cell is a count known as a................................ _____.

9. Blood is very thinly spread and then stained for studying white cells. Such a slide is described as a................ _____.

IV. LABELING

For each of the following illustrations, print the name or names of each labeled part on the numbered lines.

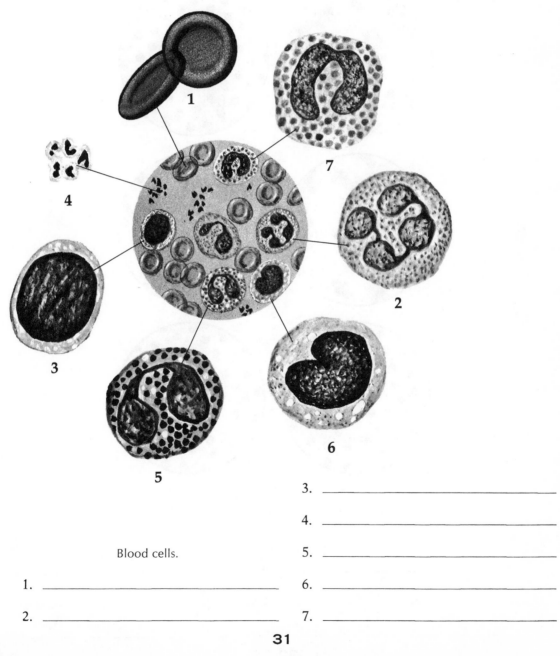

Blood cells.

3. _____

4. _____

5. _____

1. _____ 6. _____

2. _____ 7. _____

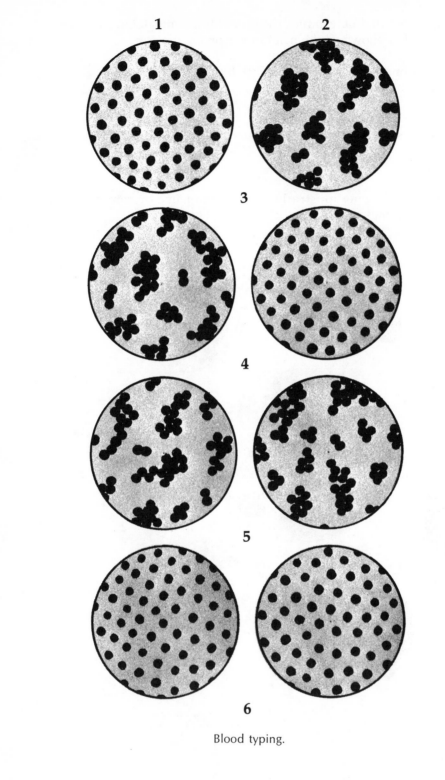

1   2

3

4

5

6

Blood typing.

1. _____ 4. _____

2. _____ 5. _____

3. _____ 6. _____

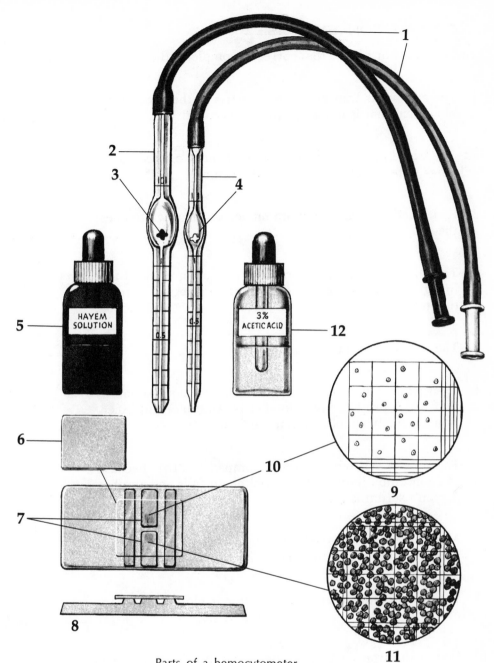

Parts of a hemocytometer.

1. _____ 7. _____

2. _____ 8. _____

3. _____ 9. _____

4. _____ 10. _____

5. _____ 11. _____

6. _____ 12. _____

V. COMPLETION EXERCISE

Print the word or phrase that correctly completes the sentence.

1. The gas that is transported to all parts of the body by the blood and that is necessary for life is called............ _____.

2. One waste product of body metabolism is carried to the lungs to be exhaled. This gas is known as.............. _____.

3. Red blood cells are far more numerous than white ones; the proportion is, in fact,............................ _____.

4. The 3-letter word that names the collection of dead and living white blood cells and bacteria in a region of infection is .... _____.

5. The erythrocytes are formed in the.................... _____.

6. Most white blood cells are formed in the same organ as that which produces red blood cells. An important exception are the lymphocytes that are formed in............ _____.

7. The process whereby cells are clumped together because of an incompatibility between red blood cells and another person's serum is called............................ _____.

8. A certain red blood cell protein is present in about 85 percent of the population. Such individuals are said to be _____.

9. The watery fluid that remains after a blood clot is removed is called .... _____.

10. One of the transport functions of the blood is the transmission of a by-product of muscle activity from the muscles to all parts of the body. This by-product is...... _____.

11. The leukocytes, or white blood cells, have as their most important function the destruction of certain disease producing organisms called .... _____.

VI. PRACTICAL APPLICATIONS

Study each discussion. Then print the appropriate word or phrase in the space provided.

Group A

1. Miss G sustained numerous deep gashes when she accidentally broke a glass shower door. One of the cuts bled copiously. In describing this type of bleeding the doctor used the word .... _____.

2. While the physician attended to the wound, the technician drew blood for typing and other studies. Miss G's blood was found to contain proteins that could be agglutinated by both anti-serums. Her blood was classified as group ... _____.

3. Among the available donors were some whose blood was found to be free of agglutinable proteins. They were classified as having blood type .... _____.

4. Further testing of Miss G's blood revealed that it lacked the Rh factor. She was therefore said to be .... _____.

5. If Miss G were to be given a transfusion of Rh positive blood, she might become sensitized to the Rh protein. In that event her blood would produce counteracting substances called .... _____.

6. Mr. B had a history of peptic ulcer. On his admission he felt weak and was having severe abdominal pain. He was hospitalized and a series of tests was begun. One of these showed a reduction in the number of red blood cells and a decrease in the hemoglobin percentage. This condition is described by a word that means an insufficiency of blood, namely .... _____.

Group B

On the medical ward there were a number of patients who required extensive blood studies.

1. A boy 7 years of age had a history of frequent fevers and a tendency to bleed easily. Physical examination revealed enlarged lymph nodes. A blood smear revealed pronounced cell changes. The percentage of each kind of white cell was determined. This is called a .... _____.

2. Further study of this patient's blood revealed a decrease in the amount of the oxygen-carrying substance known as _____.

3. Mrs. C's history included symptoms that suggested an abnormality in the use (metabolism) of the carbohydrate classified as a simple sugar and called .... _____.

4. Mr. Q complained of weakness and difficulty in walking. His red count was very low. A normal red cell count should be .... _____.

5. Mr. B, age 28, suffered from a bacterial heart disease. The laboratory technician removed a small sample of blood, took it to a centrifuge and the blood cells were separated from the liquid part, which is the .... _____.

6. Mr. K suffered from a viral infection of the liver. As a protective measure, his young son was given an injection of protein substance obtained from human plasma. This antibody, which prevents certain viral infections, has the name of .... _____.

7. Mr. B was found to lack the ability to resist his infection. He had a low white cell count, a condition that is known as _____.

8. Mr. K had been having severe blood loss from minor injuries such as the pulling of a tooth. Among other tests a blood platelet count was done. Another name for blood platelet is .... _____.

CHAPTER FOUR

Body Temperature and Its Regulation

I. OVERVIEW

Although heat is constantly being produced and lost during the course of the body's chemical activity, the body temperature is normally kept constant through its *homeostatic* mechanisms.

Homeostasis is the tendency of the body to maintain its stability despite the presence of forces that might tend to alter the situation. Through it the heart rate, blood pressure and the composition of the body fluids are also kept within the normal range.

Heat *production* is greatly increased during periods of increased muscular or glandular activity. Most heat *loss* occurs through the skin, with a smaller loss via the respiratory system and the urine and feces. The regulator responsible for keeping the temperature in a normal state regardless of heat production or heat loss is the *hypothalamus*, which transmits "messages" from the brain to the nerves so that the needed impulse is produced.

Although we usually think of the normal body temperature as being fixed at 98.6° F. (37° C.), it is more correct to speak of a *normal range*; the body temperature may vary with the time of day, the part of the body and the ingestion of food.

Abnormalities of body temperature are a valuable diagnostic tool. The presence of *fever*—an abnormally high body temperature—indicates infection most often, but may also indicate a toxic reaction, a brain injury and various other disorders. The opposite of fever is *hypothermia*—an exceedingly low body temperature—which most often comes about when the body is exposed to very low outside temperature and which can cause serious damage to the body tissues.

II. TOPICS FOR REVIEW

1. homeostasis, with examples
2. heat production within the body
3. heat loss
    a. conduction
    b. evaporation
    c. radiation
    d. convection

4. temperature regulation
5. normal temperature range
6. abnormal temperatures
   a. fever
      (1) causes
      (2) crisis
      (3) lysis
      (4) effect on phagocytosis
      (5) heat exhaustion
      (6) sunstroke (heat stroke)
   b. hypothermia
      (1) causes
      (2) effects

III. MATCHING EXERCISES

Matching only within each group, print the answer in the space provided.

Group A

| homeostasis | 36.2° to 37.6° C | extremely low |
| oxygen | extremely high | muscles and glands |

1. Body heat is produced by the combination of food products with .... _____.

2. The largest amount of heat is produced in .... _____.

3. In the condition of hypothermia, the body temperature is _____.

4. The tendency of body processes to maintain a constant state is called .... _____.

5. The normal range of body temperature is .... _____.

6. In the condition of hyperthermia, the body temperature is _____.

Group B

| skin | basal condition | blood |
| hypothalamus | tissues | subcutaneous fat |

1. The amount of heat produced by any organ depends partly on its activity and partly on its .... _____.

2. Distribution of heat throughout the body is accomplished via the .... _____.

3. The body possesses several means of ridding itself of heat; the largest part of this loss occurs through the .... _____.

4. A natural insulator against cold is the .... _____.

5. The chief heat-regulating center in the brain is the ...... ———————————.

6. The state of the body when it is at complete rest is known
   as the .... ———————————.

Group C

hypothermia	sunstroke	infection
lysis	heat exhaustion	phagocytosis
crisis		

1. Excessive loss of salt may result in the condition of ..... ———————————.

2. Failure of sweat glands to function when the body is
   exposed to high heat may result in .... ———————————.

3. Fever is most often due to .... ———————————.

4. Sometimes fever is beneficial because it steps up the
   process is which leukocytes destroy pathogens. This
   process is .... ———————————.

5. A sudden drop in temperature at the end of a period of
   fever is referred to as .... ———————————.

6. Immersion foot is an example of the kind of injury that
   may be caused by prolonged .... ———————————.

7. A gradual fall in temperature at the end of a period of
   fever is referred to as .... ———————————.

Group D

homeostasis	convection	radiation
evaporation	heat loss	heat gain
insulation	conduction	Celsius scale

1. Heat loss is accomplished in several ways. The transfer of
   heat from the body surface to the surrounding air is called ———————————.

2. Heat traveling from its source in the form of heat waves
   is called .... ———————————.

3. The freezing of water occurs at 0° and the boiling point is
   100° in the .... ———————————.

4. The amount of humidity has an effect on the rate of ..... ———————————.

5. Muscular activity, as occurs during physical exercise,
   results in .... ———————————.

6. The body has several ways of controlling its temperature.
   This regulation is one example of .... ———————————.

7. When the layer of heated air next to the body is carried away and is replaced by cooler air the process is .... _____.

8. Clothing and subcutaneous fat represent different types of _____.

9. Heat loss may occur during the conversion of a liquid or solid into a vapor, a process of .... _____.

IV. COMPLETION EXERCISE

Print the word or phrase that correctly completes the sentence.

1. While most heat loss occurs through the skin an appreciable amount is also lost in the urine, feces and via the .. _____.

2. The most important heat-regulating center is a section of the brain called the .... _____.

3. Prolonged exposure to cold may result in an abnormally low temperature, a condition named the single word .... _____.

4. A gradual drop in the temperature of a fever is known as _____.

5. Symptoms of rapid heat loss usually accompany a rapid drop in temperature which is known as .... _____.

6. During basal conditions (when the body is at rest) the organ that is believed to produce about half the body heat is the .... _____.

7. One of the most important normal ways of increasing the production of body heat is by the activity of the many organs called .... _____.

8. A fever is usually preceded by a violent attack of shivering that is best described by the one 5-letter word .... _____.

9. Because of the great increase in metabolism during a fever it is important to give the patient a diet that can best be described by the 2 words .... _____.

10. During a fever there may be considerable destruction of body tissues. Therefore, the diet should include foods that contain the nitrogenous compounds classified as .... _____.

11. Temperature control comes about in response to the heat brought to the brain by the blood as well as in response to impulses from the nerve endings in the skin called .... _____.

V. PRACTICAL APPLICATIONS

Study each discussion. Then print the appropriate word or phrase in the space provided.

A physician working in a desert area of southeastern California saw a variety of cases during the course of a day. The office nurse assisted him.

1. A 6-year-old male patient appeared apathetic and tired. His face was flushed and hot. On taking his temperature the nurse found it to be 105° F. The physician took the child's history and examined him, then instructed his mother to give the child cool sponge baths and administer the prescribed medication. The cool water sponging would aid in reducing the temperature. This is an example of cooling by means of the process of .... _____.

2. In addition to cooling the patient by means of evaporation of water on the skin, the mother was advised to use an electric fan to replace the warm air with cooler air. This is .... _____.

3. A few men working on a construction project felt faint after working only half a day. They had been perspiring profusely but had not taken the salt tablets that were placed near the drinking faucet for their use. Excessive salt loss may cause a condition called .... _____.

4. Mr. K, age 69, had been working in his garden. The day was sunny and hot, but Mr. K neglected to protect his bald head by wearing a hat. He began to feel dizzy and faint. His wife noted that his face was very flushed, and his skin appeared dry. She put him to bed at once and called the physician, who informed her that these symptoms are typical of a disorder affecting the heat-regulating sweat glands and called .... _____.

5. Mrs. K was advised to apply an ice bag to her husband's head and to give him cool sponge baths in order to reduce _____.

6. Other patients who came to see this doctor in the desert had no problems related to the hot weather. This is because the body temperature normally remains constant within quite narrow limits, an example of the concept of _____.

CHAPTER FIVE

The Skin

I. OVERVIEW

Because of its various properties, the skin comprises an *enveloping membrane*, an *organ* and a *system*. A cross section of skin would reveal its layers of *epidermis* (the outermost layer), *dermis* (the true skin where the skin glands are mainly located) and the *subcutaneous fascia* (the under-layer).

The skin serves the essential functions of *protecting* deeper tissues against drying and against invasion by harmful organisms, *regulating* body temperature and *obtaining information* from the environment. It also *excretes* salt and water in the form of perspiration. The pigment *melanin* gives the skin its color; races that have been exposed to the tropical sun for thousands of years have highly pigmented skin.

The appearance of the skin is influenced by such factors as the quantity of blood circulating in the surface blood vessels and the hemoglobin concentration. Much can be learned about the condition of the skin by observing for the presence of *discoloration*, *injury* or *eruption*. Aging, exposure to sunlight and occupational activity also have a bearing on the condition and appearance of the skin.

Being the most visible aspect of the body, the skin is the object of much quackery, and vast sums of money are spent in efforts to beautify it; good general health is, however, the most important part of skin health and beauty.

II. TOPICS FOR REVIEW

1. the skin layers
    a. epidermis
    b. dermis
    c. subcutaneous layer (superficial fascia)
2. skin glands
    a. sudoriferous glands
    b. sebaceous glands
3. functions of the skin
4. general appearance of the skin

III. MATCHING EXERCISES

Matching only within each group, print the answer in the space provided. The same answer may be used more than once.

Group A

epidermis	sebaceous glands	melanin
corium	sudoriferous	tissue fluid
integument	glands	connective
	subcutaneous layer	tissue

1. In its role as a system, the skin is called the .... _____.

2. Another term for dermis is .... _____.

3. Certain glands in the skin produce perspiration. These are
    the .... _____.

4. Since the superficial fascia is the under-the-skin layer, it
    is also known as the .... _____.

5. The oily secretion on skin and hair is produced by ...... _____.

6. Several layers of epithelial tissue form the outermost part
    of the skin, the .... _____.

7. Since the epidermis is lacking in blood vessels, nutritive
    substances reach the epidermal cells via .... _____.

8. The framework of the dermis is composed of .... _____.

9. Skin color is due largely to the presence of pigment gran-
    ules called .... _____.

Group B

trauma	receptors	dermis (or corium)
dilate	absorption	ciliary glands
pathogens	infection	

1. In its function of regulating body temperature the skin
    dissipates heat as the blood vessels enlarge or .... _____.

2. Modified sweat glands are found in the eyelid edges. These
    are known as .... _____.

3. In its role of protecting deeper tissues the skin prevents
    drying and invasion by .... _____.

4. The skin's function of obtaining information from the
    environment is due to the presence of a variety of sensory
    nerve endings. One general term for these is .... _____.

5. A framework of connective tissue and the presence of blood vessels and nerves characterize the .... —————.

6. The general term for a wound or injury to the skin or other organs is .... —————.

7. The nerve endings of the skin are located mainly in the .. —————.

8. One of the functions of the skin that is actually very minimal is .... —————.

9. Following a wound or injury of the skin, pathogens may enter and cause an .... —————.

Group C

intact	nerve endings	temperature regulator
melanin	ceruminous glands	perspiration and oil
more rapid	absorption	

1. Dilation of blood vessels brings more blood to the surface so that heat is dissipated into the air. This is one way in which the skin acts as a .... —————.

2. The skin is an able defender against invasion by pathogens as long as it remains unbroken or .... —————.

3. Exposure to sunlight causes an increase in the quantity of —————.

4. Relatively, heat loss in the child compared with that in the adult is .... —————.

5. A modification of the sweat glands is seen in the wax, or .. —————.

6. Medications are given by mouth or by injection more often than they are applied to the skin. This is because the skin has limited powers of .... —————.

7. Obtaining information about the environment is a function of the skin's .... —————.

8. The epidermal pores serve as outlets for .... —————.

IV. LABELING

Print the name or names of each labeled part on the numbered lines.

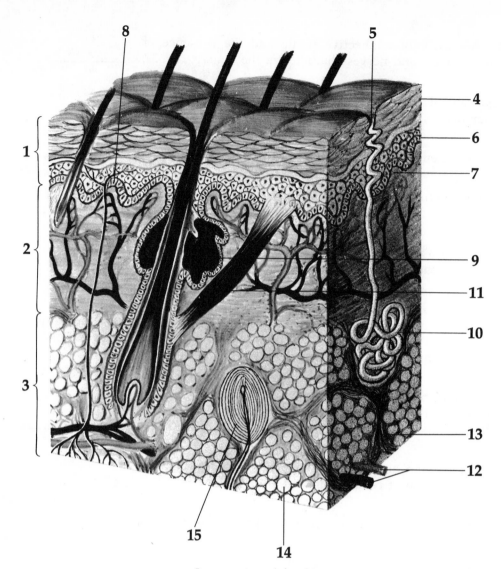

Cross section of the skin.

1. _____   9. _____

2. _____   10. _____

3. _____   11. _____

4. _____   12. _____

5. _____   13. _____

6. _____   14. _____

7. _____   15. _____

8. _____

V. COMPLETION EXERCISE

Print the word or phrase that correctly completes the sentence.

1. The outer cells of the epidermis, which are constantly being shed, are designated the horny layer, or .... _____.

2. The tissue cells of the epidermis belong to the classification called .... _____.

3. The deepest layer of the epidermis is known as the growing layer or (scientifically) the .... _____.

4. The pigment of the skin is .... _____.

5. The scientific name for a sweat gland is .... _____.

6. Oil glands are known as .... _____.

7. The scientific name that means "under the skin" is ...... _____.

8. Another name for the layer under the skin refers to its nearness to the surface and its structure. This name is ... _____.

9. The wax glands in the ear canal are called .... _____.

10. The glands on the edge of the eyelids, which are modified sweat glands, are called .... _____.

11. Areas of redness represent a type of skin injury known as _____.

12. The nails originate from the outer part of the .... _____.

13. The ceruminous glands and the ciliary glands are modified _____.

14. Hair and nails are extensions from the outer epidermis. Since they are parts that are added or appended to the skin they are called cutaneous .... _____.

VI. PRACTICAL APPLICATIONS

Study each discussion. Then print the appropriate word or phrase in the space provided.

These patients were seen in an outpatient clinic.

1. Mrs. A brought her young infant in to the clinic for information and advice. She inquired about the cheeselike covering on the baby's skin at his birth. She was told it is called .... _____.

2. Mrs. A inquired about the care of the wax in the baby's ear canal. She was instructed in the proper way to care for her baby's ears. This wax is produced by glands called _____.

3. The mother was advised to use warmer clothes for the baby during cold weather because, compared with an adult, the proportion of skin surface in an infant is considerably .... _____.

CHAPTER SIX

Bones, Joints and Muscles

I. OVERVIEW

Bones, joints and muscles are so closely interrelated in function that they are collectively called the *musculoskeletal system*. The undergirding, or framework, of a building under construction might be likened to the musculoskeletal system; in both, the superstructure is wholly dependent on the framework.

To understand how the musculoskeletal system functions, you should first visualize the *skeleton*, with its 2 main divisions, the *axial* skeleton and the *appendicular* skeleton. The bones of the skeleton come together at the *joints*, which are also classified into 2 types, the *freely movable* joints and the *inmovable* or *fixed* joints. Encasing bones and joints are the skeletal muscles. In its totality, the musculoskeletal system *protects*, *supports* and furnishes *motive* power.

II. TOPICS FOR REVIEW

1. bones
   a. structure and function
   b. divisions of the skeleton
      (1) the head
      (2) the trunk
      (3) the extremities
2. joints
   a. types
   b. structure and function
3. skeletal muscles
   a. characteristics
   b. principal skeletal muscles
      (1) the head
      (2) the neck
      (3) the chest and back
      (4) the abdomen
      (5) the lower extremities

III. MATCHING EXERCISES

Matching only within each group, print the answer in the space provided.

Group A

cartilage	red marrow	yellow marrow
periosteum	calcium salts	appendicular skeleton
axial skeleton	endosteum	vertebral column

1. The bony framework of the head and trunk forms the . . . _____.

2. Production of blood cells is carried on mainly in the . . . . . _____.

3. The combination of bones that form the framework for the extremities is called the . . . . _____.

4. The fat found inside the central cavities of long bones is _____.

5. The tough connective tissue membrane that covers bones is _____.

6. The somewhat thinner membrane that lines the central cavity of long bones is . . . . _____.

7. The pliability of the young child's bones is due to their relatively large proportion of . . . . _____.

8. A primary curve in the infant and secondary curves that develop in children are characteristic of the . . . . _____.

9. The brittleness of the old person's bones is due to their relatively large proportion of . . . . _____.

Group B

cranium	occipital bone	parietal bones
ethmoid bone	sutures	sphenoid bone
temporal bones	motion	articulation

1. The delicate spongy bone located between the eyes is called the . . . . _____.

2. At the back of the skull, and including most of the base of the skull, is situated the . . . . _____.

3. That part of the skull which encloses the brain is the . . . . _____.

4. The bat-shaped bone that extends behind the eyes and also forms part of the base of the skull is the . . . . _____.

5. The region of union of 2 or more bones is called a joint or an . . . . _____.

6. Certain joints allow for changes of position and thereby provide for .... _____.

7. The paired bones that form the larger part of the upper and side walls of the cranium are the .... _____.

8. The 2 bones that form the lower sides and part of the base of the central areas of the skull are .... _____.

9. The cranial bones join at places called .... _____.

Group C

mandible	maxillae	zygomatic bone
hyoid	nasal bones	lacrimal bone

1. At the corner of each eye is a very small bone, the .... _____.

2. The only movable bone of the skull is the .... _____.

3. Lying just below the skull proper is a U-shaped bone called the .... _____.

4. The higher part of each cheek is formed by a bone called the .... _____.

5. The 2 bones of the upper jaw are the .... _____.

6. The 2 slender bones that form much of the bridge of the nose are the .... _____.

Group D

true ribs	cervical section	floating ribs
lumbar part	vertebral column	thoracic section
rib cage	coccygeal part	foramina

1. The framework of the trunk includes the rib cage and the _____.

2. The spinal column is divided into 5 regions; the first 7 vertebrae comprise the main framework of the neck. This is the .... _____.

3. Just below the first 2 sections of the vertebral column are 5 bones that are somewhat larger than the first 19 vertebrae. These form the .... _____.

4. The second part of the vertebral column has a distinct outward curve. These 12 bones comprise the .... _____.

5. Openings or holes that extend into or through bones are called .... _____.

6. In the child the tail part of the vertebral column is made of 4 or 5 small bones that later fuse. This is the .... _____.

7. Protecting the heart and other organs as well as supportting the chest are functions of the surrounding framework called the .... _____.

8. The first 7 pairs of ribs are called the .... _____.

9. Among the false ribs, as they are called, are 2 pairs, the last 2, which are very short and do not extend to the front of the body. These are the .... _____.

Group E

patella	tibia	radius
ulna	olecranon	sesamoid

1. The upper part of the ulna forming the point of the elbow is the .... _____.

2. The medial forearm bone is the .... _____.

3. The kneecap is also called the .... _____.

4. Of the 2 bones of the leg the larger is the .... _____.

5. The forearm bone on the thumbside is the .... _____.

6. The patella is the largest of a type of bone that is encased in connective tissue. It is described as .... _____.

Group F

triceps brachii	masseter	Achilles tendon
latissimus dorsi	intrinsic	orbicularis oris
pectoralis major	trapezius	buccinator

1. The large muscle of the anterior chest region is called the _____.

2. The circular muscle of the lips is known as the .... _____.

3. Located at the angle of the jaw is an important chewing muscle called the .... _____.

4. The fleshy muscle of the cheek is the .... _____.

5. Any muscle that is located entirely within an organ or part is said to be .... _____.

6. The fibrous cord that connects the gastrocnemius to the calcaneus is called the .... ────────────.

7. A superficial muscle of the neck and upper back is called the .... ────────────.

8. The strong extensor of the elbow used in boxing is called the .... ────────────.

9. A strong back muscle that inserts on the humerus and is important in swimming is the .... ────────────.

Group G

costae	pectoral girdle	phalanges
greater trochanters	calcaneus	carpal bones
metacarpal bones	symphysis pubis	ilium
pelvic girdle	processes	foramen magnum

1. The 5 bones in the palm of each hand are the .... ────────────.

2. The largest of the tarsal bones is the heel bone or .... ────────────.

3. The 14 small bones that form the framework of the fingers on each hand are the .... ────────────.

4. In the pelvic girdle, the os coxae is divided into 3 areas. The upper wing-shaped part is the .... ────────────.

5. The bones of the wrist are the .... ────────────.

6. The clavicle and the scapula are contained in the .... ────────────.

7. The os coxae articulating with the sacrum comprise the ──────────.

8. The ribs are also designated the .... ────────────.

9. The pubic parts of the 2 ossa coxae unite to form the joint called the .... ────────────.

10. The largest opening in the skull, containing the spinal cord and related parts, is the .... ────────────.

11. The large rounded projections located at the upper and lateral portions of the femur are the .... ────────────.

12. Among the numerous prominences that serve as regions for muscle attachments are those called .... ────────────.

Group H

articular cartilage	flexion	ligaments
tendon	epimyseum	synovial membrane
voluntary muscle	aponeurosis	rotation
extension	contraction	adduction
abduction	bursa	metatarsal arch

1. The lubricating fluid inside a joint cavity is produced by the cavity lining, the .... _____.

2. Bones in the region of a joint are held together by connective tissue bands called .... _____.

3. The contacting surfaces of each joint are covered by a layer of gristle, the .... _____.

4. A bending motion that decreases the angle between 2 parts is .... _____.

5. Movement away from the midline of the body is known as _____.

6. Motion around a central axis is called .... _____.

7. When stimulated by nerve impulses, the muscle fibers become shorter and thicker; this results in muscle .... _____.

8. Muscle may be attached to bone by a cordlike structure called a .... _____.

9. Sometimes a muscle is attached to a bone by means of a sheetlike structure, an .... _____.

10. The reverse of flexion is .... _____.

11. Another term for skeletal muscle is .... _____.

12. The scissors kick of swimming is an example of .... _____.

13. The connective tissue sheath enclosing an entire muscle is the .... _____.

14. Extending crosswise under the ball of the foot is the transverse or .... _____.

15. A padlike sac that contains synovial fluid and is useful in preventing friction is a .... _____.

IV. LABELING

For each of the following illustrations, print the name or names of each labeled part on the numbered lines.

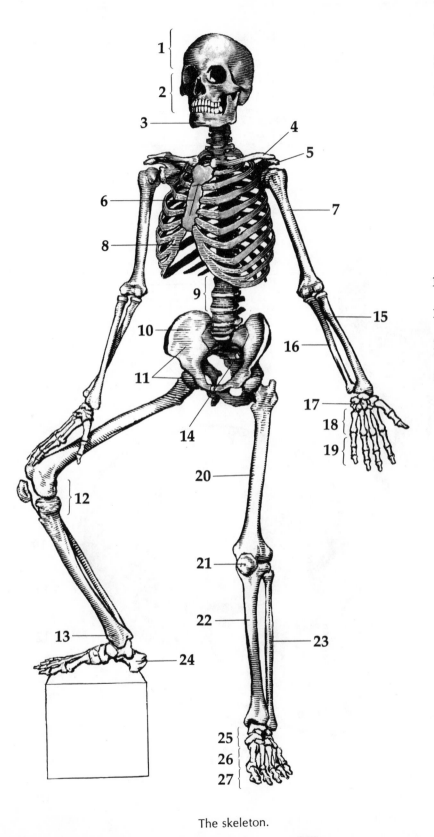

The skeleton.

1. _____
2. _____
3. _____
4. _____
5. _____
6. _____
7. _____
8. _____
9. _____
10. _____
11. _____
12. _____
13. _____
14. _____
15. _____
16. _____
17. _____
18. _____
19. _____
20. _____
21. _____
22. _____
23. _____
24. _____
25. _____
26. _____
27. _____

1. _____
2. _____
3. _____
4. _____
5. _____
6. _____

7. _____
8. _____
9. _____
10. _____
11. _____
12. _____
13. _____
14. _____
15. _____

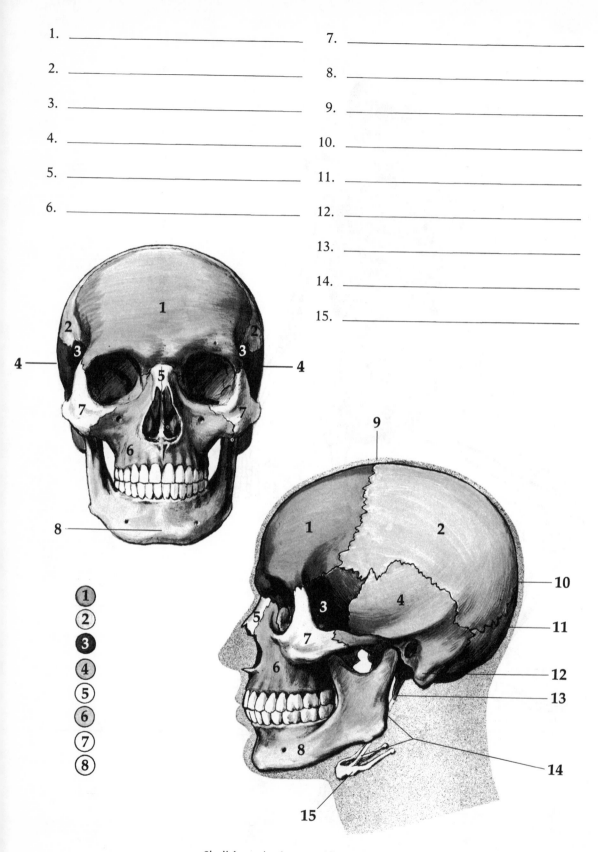

Skull from the front and from the left.

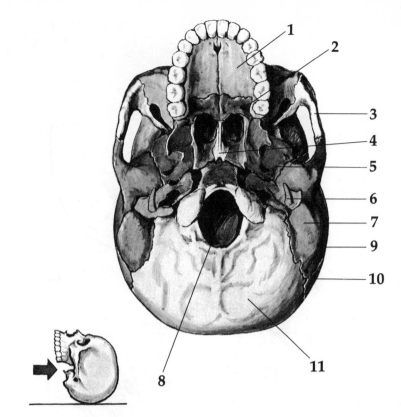

Skull from below, lower jaw removed.

1. _____

2. _____

3. _____

4. _____

5. _____

6. _____

7. _____

8. _____

9. _____

10. _____

11. _____

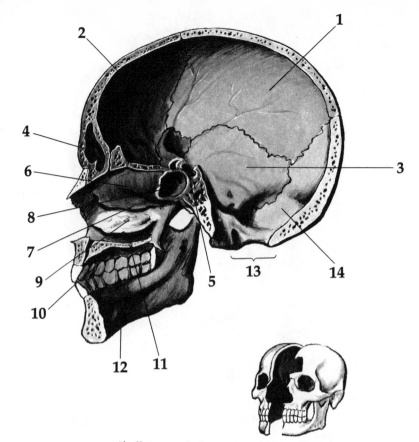

Skull, internal view.

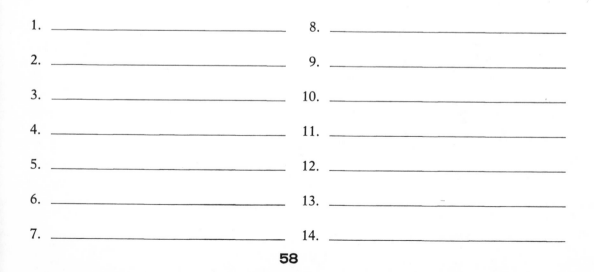

1. _____

2. _____

3. _____

4. _____

5. _____

6. _____

7. _____

8. _____

9. _____

10. _____

11. _____

12. _____

13. _____

14. _____

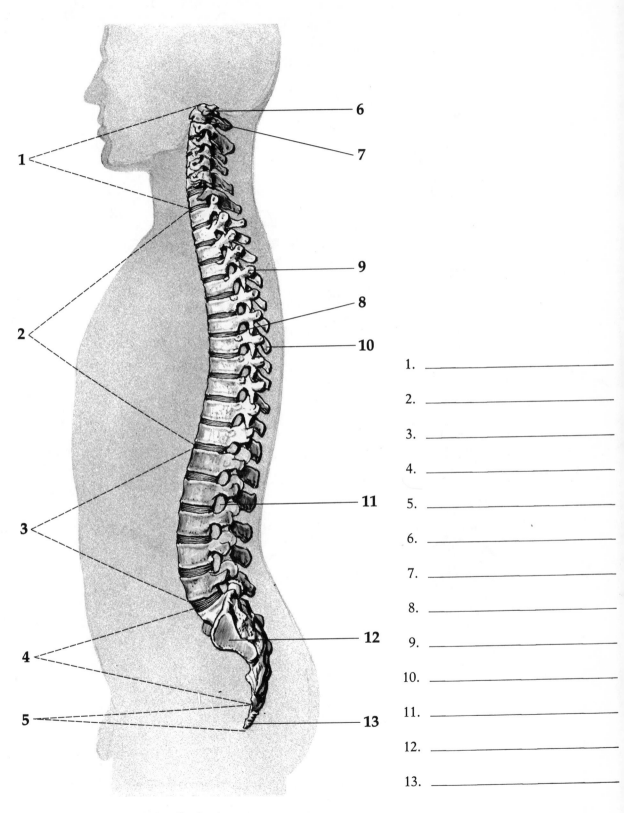

1.
2.
3.
4.
5.
6.
7.
8.
9.
10.
11.
12.
13.

1. _____

2. _____

3. _____

4. _____

5. _____

6. _____

7. _____

8. _____

9. _____

10. _____

11. _____

12. _____

13. _____

Vertebral column.

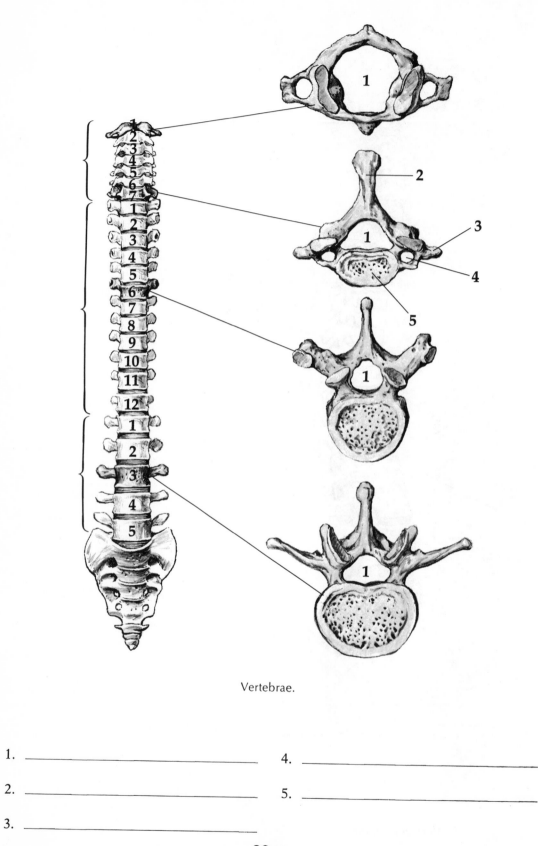

Vertebrae.

1. _____   4. _____

2. _____   5. _____

3. _____

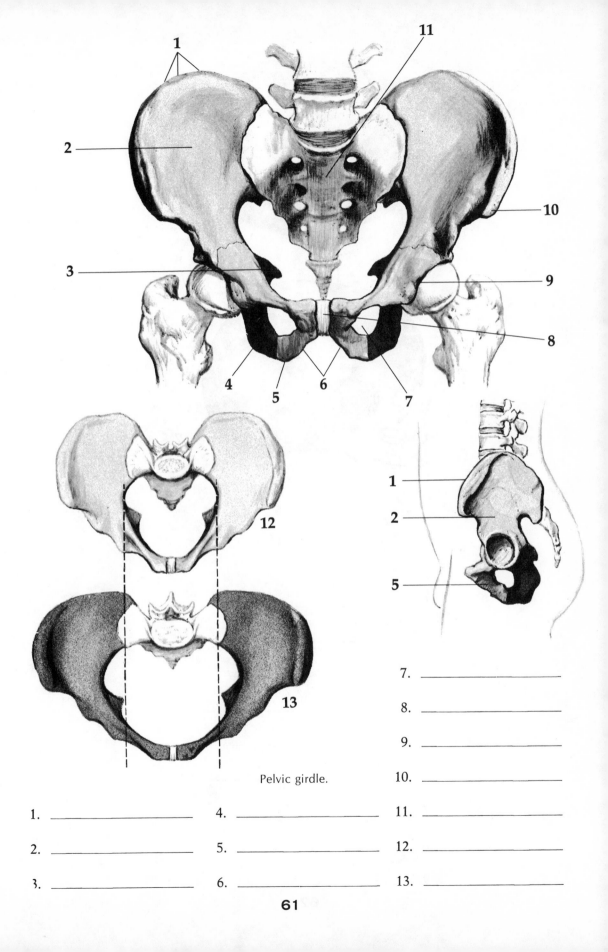

Pelvic girdle.

1. _____ 4. _____ 7. _____

2. _____ 5. _____ 8. _____

3. _____ 6. _____ 9. _____

10. _____

11. _____

12. _____

13. _____

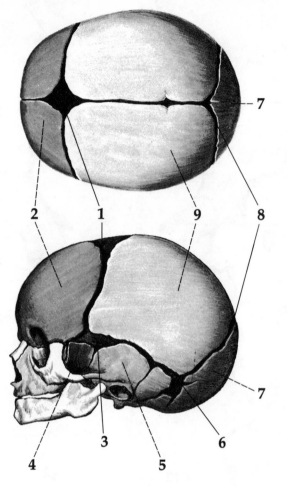

Infant skull, showing fontanels.

1. _____

2. _____

3. _____

4. _____

5. _____

6. _____

7. _____

8. _____

9. _____

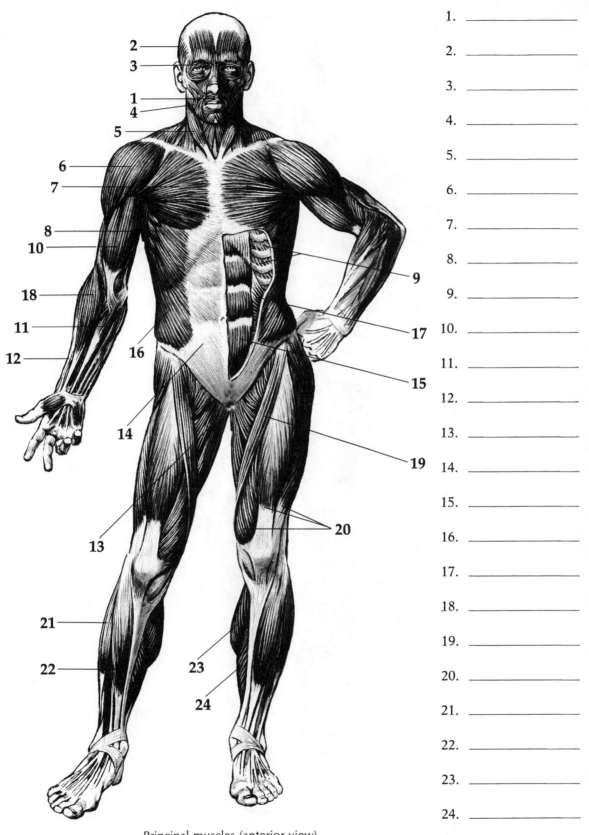

Principal muscles (anterior view).

1. _____
2. _____
3. _____
4. _____
5. _____
6. _____
7. _____
8. _____
9. _____
10. _____
11. _____
12. _____
13. _____
14. _____
15. _____
16. _____
17. _____
18. _____
19. _____
20. _____
21. _____
22. _____
23. _____
24. _____

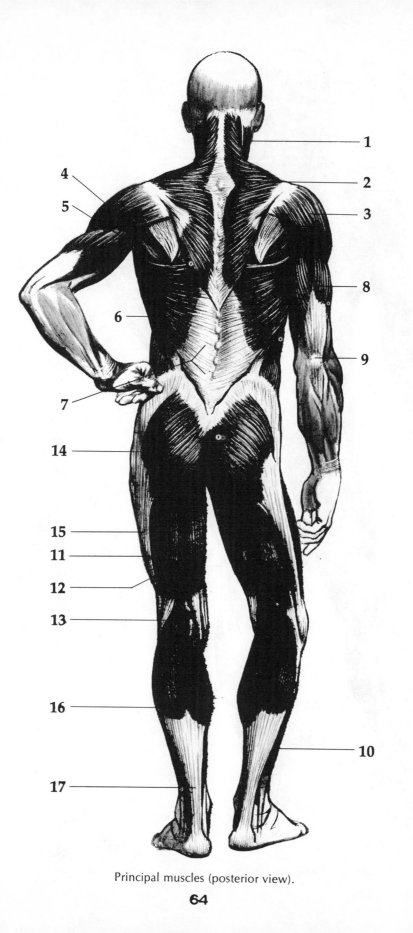

Principal muscles (posterior view).

1. _____ 10. _____

2. _____ 11. _____

3. _____ 12. _____

4. _____ 13. _____

5. _____ 14. _____

6. _____ 15. _____

7. _____ 16. _____

8. _____ 17. _____

9. _____

V. COMPLETION EXERCISE

Print the word or phrase that correctly completes the sentence.

1. Normally, muscles are in a partially contracted state, even though they are not in use at the time. This state of mild constant tension is called . . . . _____.

2. A movement is initiated by a muscle or set of muscles called the . . . . _____.

3. The movement of the prime mover is opposed by a muscle or set of muscles called the . . . . _____.

4. The muscle of the lips is the . . . . _____.

5. The flesh part of the cheek is formed by the . . . . _____.

6. There are 4 pairs of muscles of chewing. The muscles located at the angle of the jaw are called the . . . . _____.

7. A superficial muscle of the neck and upper back acts on the shoulder. This muscle is the . . . . _____.

8. Injury to the sternocleidomastoid may result in a disorder called wryneck, or . . . . _____.

9. The largest forearm extensor is the . . . . _____.

10. The muscular partition between the thoracic and abdominal cavities is the .... _____.

11. The large fleshy muscle of the buttocks which extends the hip is the .... _____.

12. The chief muscle of the calf is the toe dancer's muscle, named the .... _____.

13. The largest tendon in the body is the .... _____.

14. Bending the ankle so that the sole of the foot is facing outward, away from the body, is called .... _____.

15. The muscle attachment that is usually relatively fixed is called its .... _____.

16. Muscles contract and exert power on the more movable attachment, its .... _____.

17. Muscle fibers shorten and this contraction results in the actual movement of the muscle, its .... _____.

18. The endings of the motor nerve fibers are called motor end plates or .... _____.

VI. PRACTICAL APPLICATIONS

Study each discussion. Then print the appropriate word or phrase in the space provided.

Group A

A group of high school seniors were involved in a serious traffic accident on their way home from the "prom."

1. There was a pronounced swelling of the upper right side of Mary's head. X-ray films showed a fracture of the largest skull bone, the .... _____.

2. Mary also suffered an injury to one of the 2 large bones of the pelvic girdle. This bone articulates with the sacrum and is named the .... _____.

3. John suffered multiple injuries to his left lower extremity. Protruding through the skin was a splintered portion of the longest bone in the body, the .... _____.

4. The muscle on the front of the thigh was involved in John's injury. This large 4-part muscle is the .... _____.

5. Susan thought her injuries were the least serious, so she walked several blocks to find help. Then she noticed that her right knee was not functioning normally. Examination revealed a fractured kneecap. Another name for the knee-cap is .... ———————————.

6. Harry, the driver of the car, was forcibly thrown against the steering wheel. He suffered fractures of the sixth and seventh ribs, which are included among the .... ———————————.

7. Harry's chest was cut by flying glass. The largest anterior muscle in this area is the .... ———————————.

Group B

Patients that were visiting a doctor's office included:

1. Mr. B, age 58, who complained of acute pain and swelling of his right great toe. X-ray films showed involvement of the joints of the toe. The framework of the toes is made up of bones that are called .... ———————————.

2. Mr. B was also examined for lower back discomfort. X-rays were taken of the portion just above the sacrum, a section that is called the .... ———————————.

3. Mrs. C, age 36, complained of swelling and pain involving the fingers and hands. Examination showed that there was an apparent inflammation of the joint cavity lining. This is the .... ———————————.

4. Mrs. C also had indications of damage to the normally smooth gristle that covers the joint surface. The scientific name for this layer is .... ———————————.

CHAPTER SEVEN

The Brain, the Spinal Cord and the Nerves

I. OVERVIEW

The nervous system is the body's *coordinating system*, receiving, sorting out and responding to both internal and external stimuli. The nervous system as a whole is divided into the *central* nervous system, consisting of the brain and the spinal cord, and the *peripheral* nervous system, made up of the cranial and the spinal nerves. The central nervous system includes the *speech centers*, which are essential to one's ability to hear, see, speak and write, as well as certain parts that control such vital functions as respiration, heart rate and body balance. The peripheral nervous system controls both the *general* and the *special* sense impulses. Certain nerves and centers in this system are grouped together as the *autonomic* nervous system, because they control activities that go on more or less automatically, regulating the actions of glands, smooth muscle and the heart.

The whole of the central nervous system functions via billions of *neurons*, the structural units of the nervous system, each of which is composed of a *cell body* and *nerve fibers* that carry impulses to and away from the *cell body*.

II. TOPICS FOR REVIEW

1. structure and function of nervous system as a whole
   a. the nerve cell
   b. the nerve
2. divisions
   a. central nervous system
   b. peripheral nervous system
   c. autonomic nervous system
3. central nervous system
   a. the brain

(1) main parts
(2) cerebral hemispheres
(3) cerebral cortex
(4) speech
(5) interbrain, midbrain, cerebellum, pons, medulla oblongata
(6) ventricles
(7) brain waves
b. spinal cord
(1) location
(2) structure
(3) function
(4) spinal puncture
c. coverings of brain and spinal cord
d. cerebrospinal fluid
4. peripheral nervous system
a. cranial nerves
b. spinal nerves
5. autonomic nervous system
a. parts
b. functions

III. MATCHING EXERCISES

Matching only within each group, print the answer in the space provided.

Group A

brain and spinal cord	cerebral hemispheres
autonomic nervous system	nerve
nerve fibers	brain stem
stimuli	neuron
peripheral nervous system	coordinator

1. In relation to the parts and organs of the body, the nervous system functions as .... —————————.

2. The internal and external changes that affect the nervous system are called .... —————————.

3. For study purposes, the entire nervous system has been divided into 2 large systems. One of these, the central nervous system, is composed of the .... —————————.

4. The other large nervous system is the .... —————————.

5. The sympathetic and parasympathetic nervous systems are the 2 functionally opposing parts of the .... —————————.

6. The neuron is composed of a cell body with the addition of threadlike cytoplasmic projections, the .... —————————.

7. The cerebrum is the largest part of the brain. It is divided into right and left parts called the .... _____.

8. The midbrain, pons and medulla oblongata form the .... _____.

9. The basic nerve cell, including the cell body and its projections, is the .... _____.

10. The peripheral nerves that regulate activities going on more or less automatically are grouped together as the .. _____.

11. Impulses are conducted from one place to another by the bundle of nerve fibers, the .... _____.

Group B

sulci   mixed nerves   afferent nerves
receptor   convolutions   efferent nerves
cerebral cortex   lateral ventricles

1. The place where the stimulus is received is called the end organ or .... _____.

2. Impulses must be carried to and away from the brain and spinal cord. Those that conduct impulses to the brain and cord are grouped together as the .... _____.

3. The nerve fibers that carry impulses away from the brain and cord to muscles and glands form the .... _____.

4. All thought, association and judgment take place in the .. _____.

5. Among the important landmarks in the cerebral hemispheres are several fissures, or .... _____.

6. The fissures serve to separate the gray matter in folds forming elevated portions known as .... _____.

7. Cerebrospinal fluid fills spaces within the hemispheres. These are the .... _____.

8. Combinations of afferent and efferent fibers form .... _____.

Group C

internal capsule   temporal lobe   thalamus
lobes   motor cortex   parietal lobe
occipital lobe

1. The cerebral cortex of each hemisphere is divided into regions each of which regulates certain types of functions. These areas are called .... _____.

71

2. In each frontal lobe is an area that controls voluntary muscles. This is the .... _____ .

3. Pain, touch and temperature are interpreted in the sensory area which is contained in the .... _____ .

4. Impulses received by the ear are interpreted in the auditory center, which is located in the .... _____ .

5. Messages from the retina are interpreted in the visual area of the .... _____ .

6. The white matter of the cerebral cortex consists of collections of nerve fibers, one group of which is particularly vulnerable to injury. This area is the .... _____ .

7. Two masses of gray matter which are located in the diencephalon act as relay centers monitoring sensory stimuli. These 2 masses constitute the .... _____ .

Group D

ventricles	corpora quadrigemina	myelinated nerve
muscles of respiration	encephalogram (or	fibers
cerebrum	ventriculogram)	medulla oblongata
electroencephalograph	cerebellum	hypothalamus
blood pressure	diencephalon	

1. By cutting into the central section of the brain, one can see the interbrain, or .... _____ .

2. The vermis and the 2 lateral hemispheres at the sides form the .... _____ .

3. The 2 cerebral hemispheres form much of the largest part of the brain, the .... _____ .

4. The 4 fluid-filled spaces within the brain are called ...... _____ .

5. The respiratory, cardiac and vasomotor centers are found in the .... _____ .

6. The vasomotor center affects muscles in the blood vessel walls and thus influences .... _____ .

7. To aid in the diagnosis of tumors and other brain disorders an x-ray picture is used. It is called an .... _____ .

8. The measurable electric currents produced by the activity of the brain cells are recorded by the .... _____ .

9. Body temperature, sleep, the heart beat and water balance are among the vital body functions regulated by the _____.

10. The respiratory center exerts control over the .... _____.

11. The relay centers for eye and ear reflexes are located in the midbrain. They are the 4 .... _____.

12. The pons is white in color because it is made largely of ... _____.

Group E

afferent nerves nerve cell bodies efferent nerves
receptor   effector   nerve fibers

1. The internal section of the spinal cord is composed of gray matter consisting of the .... _____.

2. Surrounding the gray part is a larger area made up of white matter, the .... _____.

3. The spinal cord has several essential functions. One of these is to conduct sensory impulses upward to the brain in tracts within the cord. These impulses are brought to the cord by .... _____.

4. The spinal cord also functions as a pathway for conducting motor impulses from the brain downward in descending tracts. These motor impulses leave the cord via ..... _____.

5. The reflex pathway begins with the part of a sensory neuron called a .... _____.

6. The sensory neuron conducts an impulse to a central neuron which then transfers it to a motor neuron. This typical reflex pathway terminates in a gland or a muscle with an.. _____.

Group F

dura mater arachnoid membrane pia mater
arachnoid villi choroid plexuses   subarachnoid space

1. The innermost layer of the meninges, the delicate connective tissue membrane in which there are many blood vessels, is the .... _____.

2. The weblike middle meningeal layer is the .... _____.

3. The outermost meningeal layer, which is the thickest and toughest, is also made of connective tissue. It is the ..... _____.

4. Normally the cerebrospinal fluid helps protect the brain and spinal cord against shock. This fluid is formed inside the brain ventricles by the .... _____.

5. Normally, the fluid flows freely from ventricle to ventricle and finally out into the .... _____.

6. The fluid is returned to the blood in the venous sinuses through the projections called .... _____.

Group G

visual area	auditory speech	visual speech
written speech	center	center
center	sensory area	

1. Pain, touch, temperature, size and shape are interpreted in the parietal lobe, in a section called the .... _____.

2. The understanding of words takes place with the development of a temporal lobe area known as the .... _____.

3. Messages from the retina are interpreted in the region of the occipital lobe known as the .... _____.

4. The ability to read with understanding comes with the development of the .... _____.

5. The ability to write words, which usually is a late phase in a person's total language comprehension, is a function of the .... _____.

Group H

ganglion	sensory cell ganglia	roots
cervical plexus	plexuses	brachial plexus

1. Each spinal nerve is attached to the spinal cord by combinations of nerve fibers called .... _____.

2. The small masses of nerve cell bodies attached to each dorsal root are the .... _____.

3. A collection of nerve cell bodies usually found outside the central nervous system is a .... _____.

4. A short distance away from the spinal cord each spinal nerve branches into 2 divisions; the branches of the larger division interlace to form .... ————————.

5. The shoulder, the arm, the wrist and the hand are supplied by branches from the .... ————————.

6. Motor impulses to the neck muscles are supplied by the .. ————————.

Group I

parasympathetic nervous system   oculomotor nerve
sympathetic nervous system   trigeminal nerve
hypoglossal nerve   acoustic nerve
olfactory nerve   optic nerve
vagus nerve   facial nerve

1. Recall the functions of the autonomic nervous system. The part that acts to prepare the body for emergency situations is the .... ————————.

2. The part of the autonomic nervous system that aids the digestive process is the .... ————————.

3. Impulses controlling tongue muscles are carried by the .. ————————.

4. General sense impulses to the face and head are carried through the 3 branches of the .... ————————.

5. Sense fibers for hearing are contained within the .... ————————.

6. The muscles of facial expression are supplied by branches of the .... ————————.

7. The nerve that carries smell impulses to the brain is the .. ————————.

8. The contraction of most eye muscles is controlled by the .. ————————.

9. The visual nerve is called the .... ————————.

10. Most of the organs in the thoracic and abdominal cavities are supplied by the .... ————————.

IV. LABELING

For each of the following illustrations, print the name or names of each labeled part on the numbered lines.

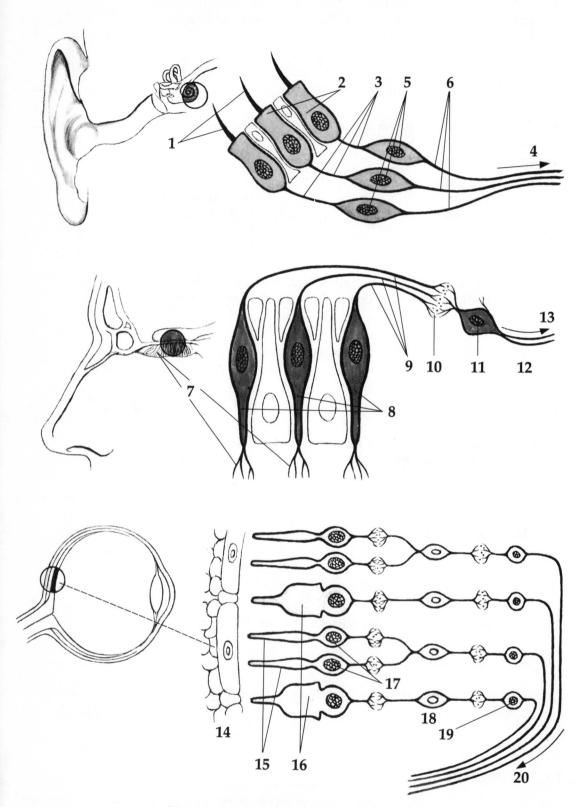

Diagram of neurons for receiving special senses.

76

1. _____

2. _____

3. _____

4. _____

5. _____

6. _____

7. _____

8. _____

9. _____

10. _____

11. _____

12. _____

13. _____

14. _____

15. _____

16. _____

17. _____

18. _____

19. _____

20. _____

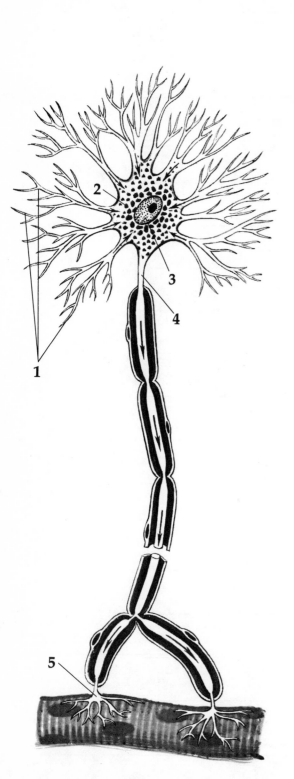

1. _____

2. _____

3. _____

4. _____

5. _____

Diagram of a motor neuron.

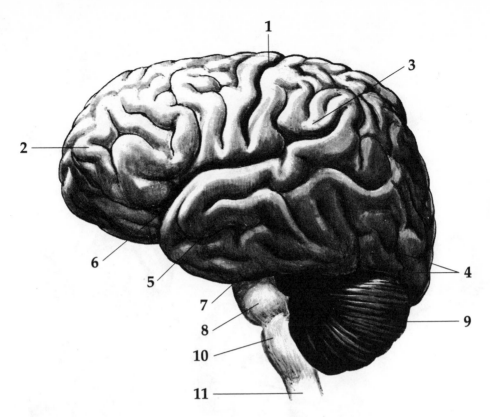

The external surface of the brain.

1. _____

2. _____

3. _____

4. _____

5. _____

6. _____

7. _____

8. _____

9. _____

10. _____

11. _____

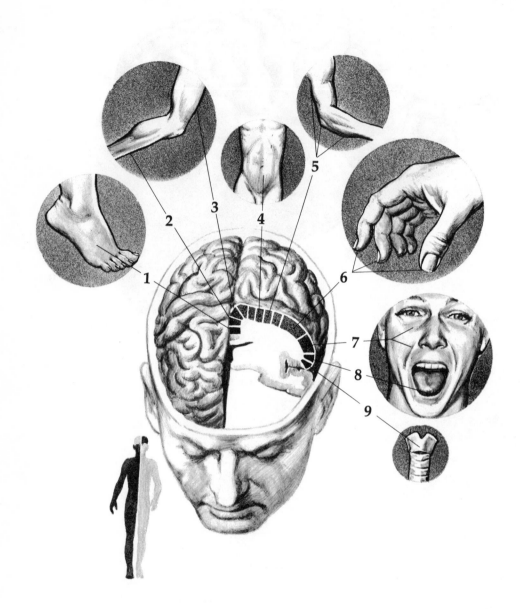

The motor area of the left cerebral hemisphere.

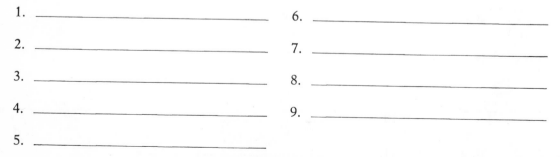

1. _____   6. _____

2. _____   7. _____

3. _____   8. _____

4. _____   9. _____

5. _____

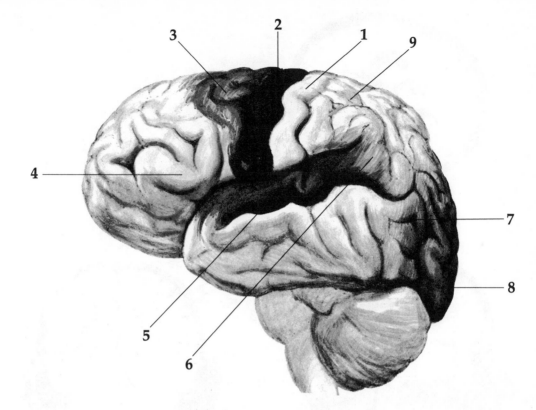

The functional areas of the cerebrum.

1. _____

2. _____

3. _____

4. _____

5. _____

6. _____

7. _____

8. _____

9. _____

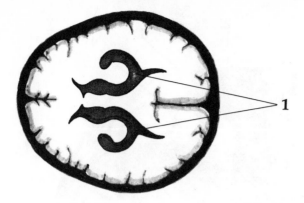

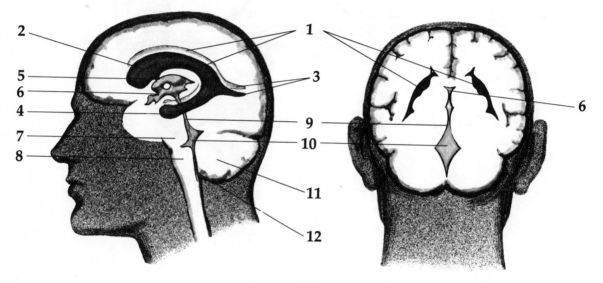

Brain ventricles.

1. _____   7. _____

2. _____   8. _____

3. _____   9. _____

4. _____   10. _____

5. _____   11. _____

6. _____   12. _____

1. _____   3. _____

2. _____   4. _____

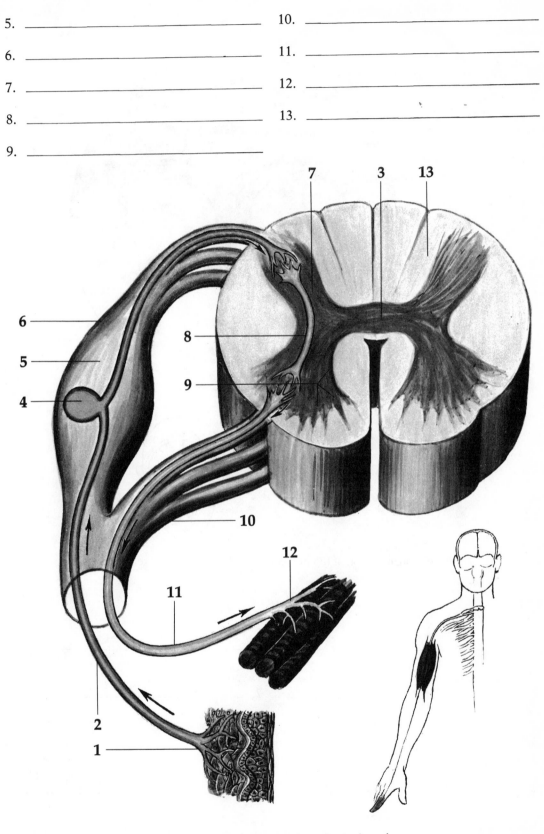

Reflex arc and cross section of spinal cord.

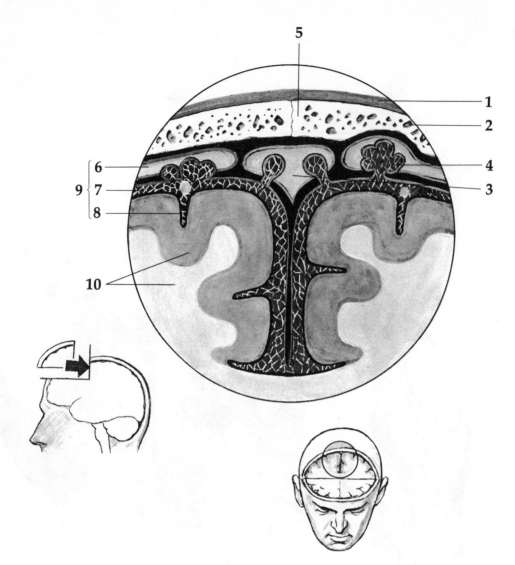

Frontal (coronal) section of top of head to show meninges and related parts.

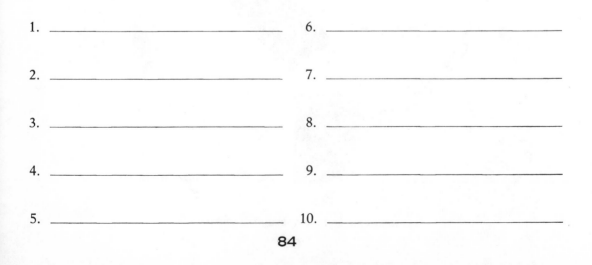

1. _____   6. _____

2. _____   7. _____

3. _____   8. _____

4. _____   9. _____

5. _____   10. _____

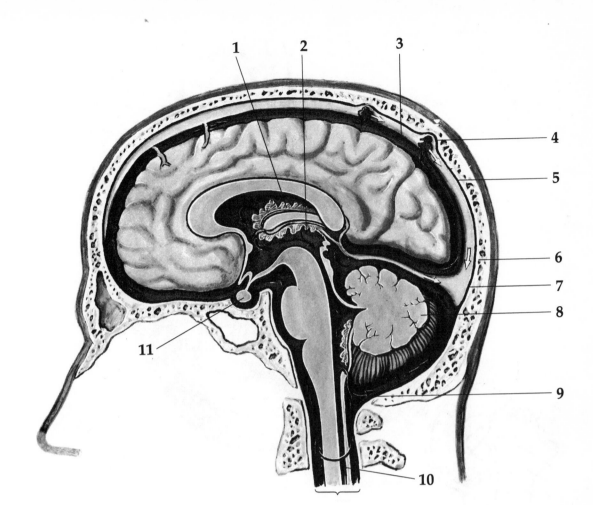

Flow of cerebrospinal fluid.

1. _____	7. _____
2. _____	8. _____
3. _____	9. _____
4. _____	10. _____
5. _____	11. _____
6. _____	

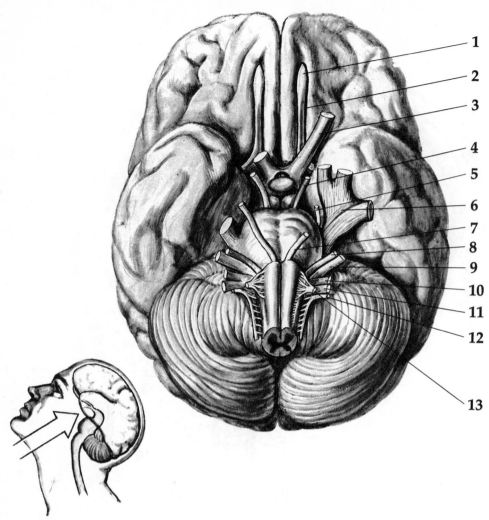

Base of brain, showing cranial nerves.

1. _____   8. _____

2. _____   9. _____

3. _____   10. _____

4. _____   11. _____

5. _____   12. _____

6. _____   13. _____

7. _____

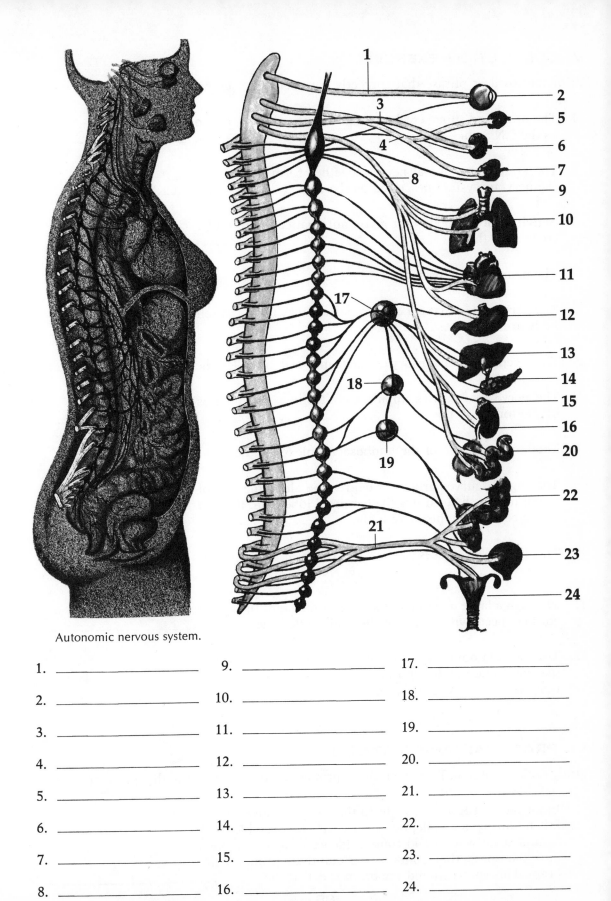

Autonomic nervous system.

1. _____   9. _____   17. _____

2. _____   10. _____   18. _____

3. _____   11. _____   19. _____

4. _____   12. _____   20. _____

5. _____   13. _____   21. _____

6. _____   14. _____   22. _____

7. _____   15. _____   23. _____

8. _____   16. _____   24. _____

V. COMPLETION EXERCISE

Print the word or phrase that correctly completes the sentence.

1. The brain and spinal cord together are usually referred to as the C.N.S., or .... _____.

2. The cranial and spinal nerves together form the part of the nervous system described as the .... _____.

3. Activities of the body that go on automatically are under control of the .... _____.

4. The neuron is the basic .... _____.

5. The nerve fibers that conduct impulses away from the cell body are the .... _____.

6. The receptor of the sensory dendrite may also be called the _____.

7. The nerve that carries impulses controlling the muscles of the tongue is the .... _____.

8. The largest branch of the lumbosacral plexus is the ..... _____.

9. The slightly curved groove or depression along the side of the brain which separates the temporal lobe from the rest of the cerebral hemisphere is the .... _____.

10. The fluid-filled spaces within the cerebral hemispheres are the .... _____.

11. Through a large opening in the base of the skull the spinal cord connects with a part of the brain called the .... _____.

12. The medulla oblongata contains collections of cell bodies that have to do with important vital centers. These collections are known as centers, or .... _____.

VI. PRACTICAL APPLICATIONS

Study each discussion. Then print the appropriate word or phrase in the space provided.

1. Eight-year-old K was brought to the clinic because he had fallen during an epileptic seizure. There was bleeding from a scalp wound and some evidence of a hemorrhage into the space that contains a fluid that cushions the brain. This space around the brain is called the .... _____.

2. Mrs. M's son found his mother lying unconscious on the floor. Mrs. M was 67 years old and had a history of high blood pressure. She was admitted to the intensive care unit. She was unable to speak or write, or to understand written or spoken language. She had injury to the speech centers which are located in the part of the brain called the _____.

3. Mr. H, age 42, had been suffering for several weeks from persistent intractable headaches. An x-ray study of the brain was ordered. Such an x-ray is called an . . . . _____.

4. Some of the fluid was removed from the ventricles in Mr. H's brain and replaced with air, as part of the diagnostic study. This fluid is the . . . . _____.

5. As a result of the various studies done in Mr. H's case it was determined that a tumor was present in the left lateral ventricle. Surrounding the left ventricle is the . . . . _____.

6. Miss S's symptoms included paralysis and various motor disturbances. The diagnosis of myelitis, or inflammation of the spinal cord was made. This nerve cord is located in a space called the . . . . _____.

CHAPTER EIGHT

The Sensory System

I. OVERVIEW

Through the functioning of the *sensory receptors* we are made aware of all changes taking place both internally (within the body) and externally (outside the body). The *special* senses—so-called because the receptors are limited to a relatively small area of the body—include the visual sense, the hearing sense, the senses of taste and smell, those of hunger and appetite and the sense of thirst. The *general* senses are scattered throughout the body; they have to do with pressure, temperature, pain, touch and position.

II. TOPICS FOR REVIEW

1. the eye
     a. protective structures of eyeball
     b. coats of eyeball
     c. pathway of light rays
     d. receptors (sensory end organs)
     e. extrinsic eyeball muscles
     f. intrinsic eyeball muscles
     g. nerves
     h. lacrimal apparatus
2. the ear
     a. external ear
     b. middle ear
     c. inner ear
3. other organs of special sense
     a. taste
     b. smell
     c. hunger and appetite
     d. thirst

4. general sense organs
    a. pressure
    b. temperature
    c. touch
    d. pain
    e. position

III. MATCHING EXERCISES

Matching only within each group, print the answer in the space provided. The same answer may be used more than once.

Group A

cornea   aqueous humor   transparent refracting parts
accommodation   rods and cones   choroid coat
retina   vitreous body

1. The innermost coat of the eyeball, the nerve tissue layer, includes the end organs for the sense of vision. This structure is the .... _____.

2. The pigmented middle tunic of the eyeball is the vascular _____.

3. Light rays pass through a series of transparent eye parts. The outermost of these is the .... _____.

4. The watery fluid that fills much of the eyeball in front of the crystalline lens and also helps to maintain the slight curve in the cornea is the .... _____.

5. The spherical shape of the eyeball is maintained by a jelly-like material located behind the crystalline lens. This is the _____.

6. The receptors for the sense of vision are called the ...... _____.

7. The elasticity of the lens enables it to become thicker and bend the light rays as necessary. This process is .... _____.

8. The media of the eye may be described as the .... _____.

9. Bulging forward slightly is the "window," or .... _____.

Group B

iris   pupil   ciliary body
media   sclera   receptors
optic disk   conjunctiva

92

1. The opaque outermost layer of the eyeball is made of firm, tough connective tissue. This coat is the .... ——————————.

2. The central opening in the iris contracts or dilates according to need. This opening is the .... ——————————.

3. The crystalline lens is one of the transparent refracting parts of the eye. Collectively they are called .... ——————————.

4. The rods and cones of the retina are the visual end organs or .... ——————————.

5. The membrane that lines the eyelids is the .... ——————————.

6. The region of connection between the optic nerve and the eyeball is lacking in rods and cones and is commonly called the blind spot. Another term for this is .... ——————————.

7. The shape of the lens is altered by the muscle of the ..... ——————————.

8. The pupil is the central opening in the colored part of the eye, the .... ——————————.

Group C

fovea centralis	iris	intrinsic
sphincter	lacrimal gland	conjunctiva
extrinsic	refraction	position and balance

1. The muscles that are attached to bones of the orbit and to the sclera are located outside the eyeball and are described as .... ——————————.

2. When a light is flashed in the eye the pupil is reduced in size due to the contraction of an iris muscle, the .... ——————————.

3. The amount of light entering the eye is controlled by the ——————————.

4. The process of bending which makes it possible for light from a large area to be focused on a small surface is known as .... ——————————.

5. Tears serve an important protective function for the eye. They are produced by the .... ——————————.

6. The clearest point of vision is a depressed area in the retina, the .... ——————————.

7. From the muscles, the joints and the semicircular canals come the sense of .... _____ .

8. Separating the front of the eye from the eyeball proper is a sac lined with an epithelial membrane called the ...... _____ .

9. The muscles of the iris and ciliary body are located entirely within the eyeball and so are described as .... _____ .

Group D

oval window	external auditory canal	eustachian tube
ossicles	endolymph	tympanic membrane
perilymph	pinna	mastoid air cells

1. Located at the end of the auditory canal is the eardrum, or _____ .

2. The 3 small bones within the middle ear cavity are the ... _____ .

3. The spaces within the temporal bone which connect with the middle ear cavity through an opening are called the... _____ .

4. Sound waves are conducted to the fluid of the internal ear by vibrations of the membrane that covers the .... _____ .

5. Air is brought to the middle ear cavity by means of the auditory tube which is also called the .... _____ .

6. The fluid of the inner ear contained within the bony labyrinth and surrounding the membranous labyrinth is called .... _____ .

7. The fluid contained within the membranous labyrinth is called .... _____ .

8. Sound waves enter the .... _____ .

9. Another name for the projecting part, or auricle, of the ear is the .... _____ .

Group E

optic nerve	vestibule	ophthalmic nerve
oculomotor nerve	cochlear duct	temporal

1. The organ of hearing is made up of receptors located in the .... _____ .

2. The entrance area that communicates with the cochlea and that is next to the oval window is the .... ———————.

3. Visual impulses received by the rods and cones of the retina are carried to the brain by the .... ———————.

4. Impulses of pain, touch and temperature are carried to the brain by a branch of the fifth cranial nerve, the ...... ———————.

5. The largest cranial nerve carrying motor fibers to the eyeball muscles is the .... ———————.

6. Most of the parts of the ear are located within one of the major bones of the skull, namely the .... ———————.

Group F

taste buds   ceruminous   olfactory
adaptation   polydipsia   pressure

1. The sense of taste involves 2 cranial nerves as well as receptors known as .... ———————.

2. Excessive thirst, as may occur in certain illnesses including diabetes, is referred to as .... ———————.

3. Among the general senses is that concerned with deep sensibility, commonly called the sense of .... ———————.

4. In the case of many sensory receptors, including those for temperature, the receptors adjust themselves so that one does not feel the sensation so acutely if the original stimulus is continued. Such an adjustment to the environment is called .... ———————.

5. The wax glands located in the external auditory canal are described as .... ———————.

6. The pathway for impulses from smell receptors is the first cranial nerve, the .... ———————.

IV. LABELING

For each of the following illustrations, print the name or names of each labeled part on the numbered lines.

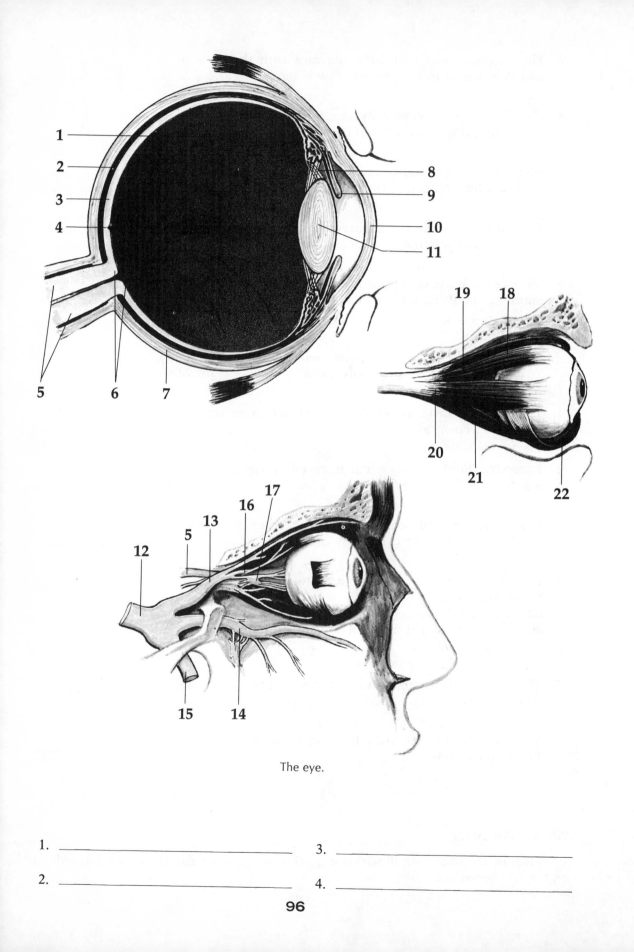

The eye.

1. _____   3. _____

2. _____   4. _____

5. _____
6. _____
7. _____
8. _____
9. _____
10. _____
11. _____
12. _____
13. _____

14. _____
15. _____
16. _____
17. _____
18. _____
19. _____
20. _____
21. _____
22. _____

Lacrimal apparatus.

1. _____
2. _____
3. _____
4. _____

5. _____
6. _____
7. _____
8. _____

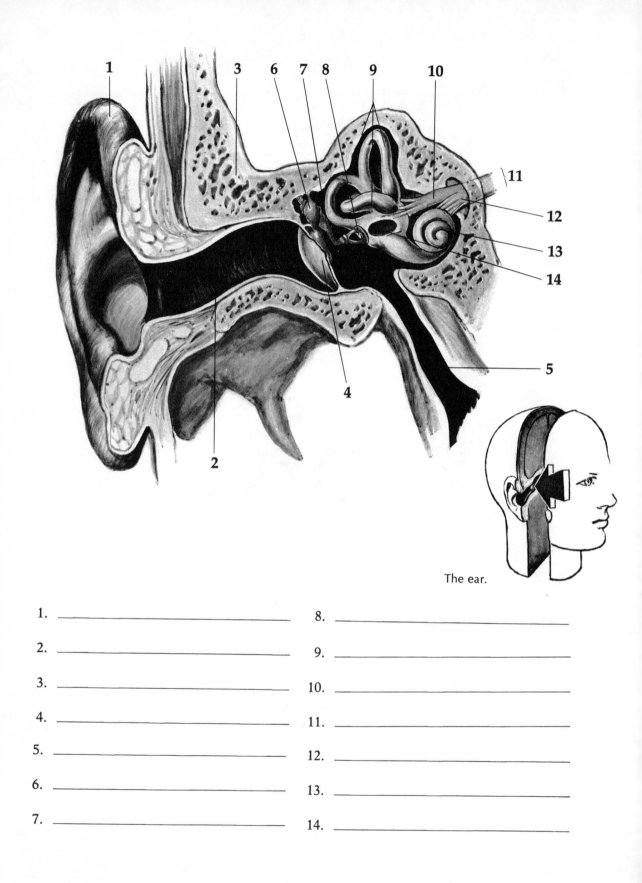

The ear.

1. _____

2. _____

3. _____

4. _____

5. _____

6. _____

7. _____

8. _____

9. _____

10. _____

11. _____

12. _____

13. _____

14. _____

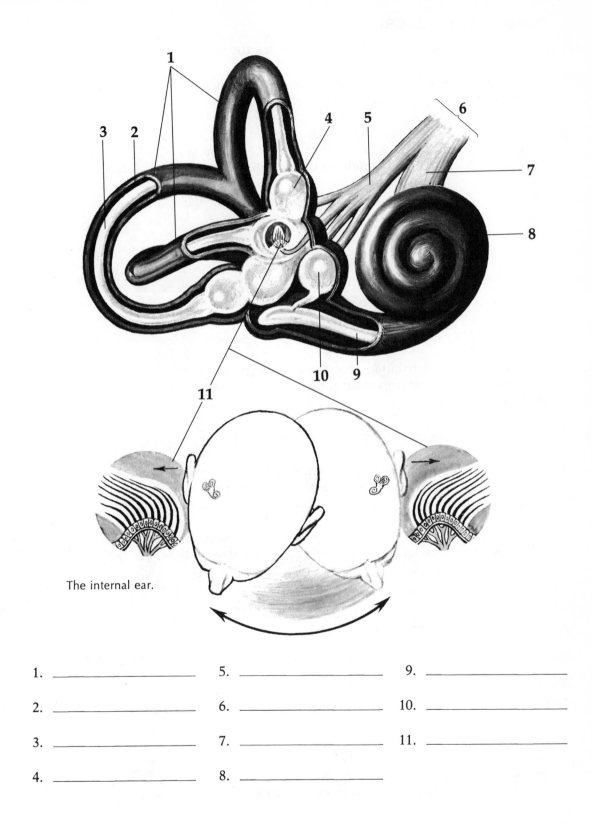

The internal ear.

1. _____   5. _____   9. _____

2. _____   6. _____   10. _____

3. _____   7. _____   11. _____

4. _____   8. _____

V. COMPLETION EXERCISE

Print the word or phrase that correctly completes the sentence.

1. The nerve fibers of the vestibular and cochlear nerves join to form the auditory nerve, more often called the ... _____.

2. The inner ear spaces contain fluids involved in the transmission of sound waves. The one that is inside the membranous cochlea and that stimulates the receptors is the .. _____.

3. The taste receptors of the tongue are located along the edges of small depressed areas, or .... _____.

4. The nerves involved in the sense of taste are the facial and the .... _____.

5. Pain that is felt in an outer part of the body such as the skin, yet that originates internally near the area where it is felt is called .... _____.

6. The very widely distributed free nerve endings are the receptors for the most important protective sense, namely that for .... _____.

7. The tactile corpuscles are the receptors for the sense of .. _____.

8. The nerve endings that relay impulses which aid in judging position and changes in location of parts with respect to each other are the .... _____.

9. The sense of position is partially governed by several structures in the internal ear, including 2 small sacs and the 3 membranous .... _____.

VI. PRACTICAL APPLICATIONS

Study each discussion. Then print the appropriate word or phrase in the space provided.

Group A

While observing in the emergency ward the student nurse noted the following cases.

1. Ten-year-old K had been riding his bicycle while he threw glass bottles to the sidewalk. A fragment of glass flew into one eye. Examination at the hospital showed that there was a cut in the transparent window of the eye, the ..... _____.

2. On further examination of K, the colored part of the eye was seen to protrude from the wound. This part is the ... _____.

3. K's treatment included antiseptics, anesthetics and suturing of the wound. Medication was instilled in the saclike structure at the front of the eyeball. This sac is lined with a thin epithelial membrane, the .... _____.

4. A construction worker, Mr. J, was admitted because of an accident in which a piece of steel penetrated his eyeball and caused such an extensive wound that material from the inside of the eyeball oozed out. Mr. J tried to relieve the pain by forcing the jellylike material out through the wound at the front and side of his eyeball. This matter, which maintains the shape of the eyeball, is called the ... _____.

Group B

An ear, nose and throat specialist treated the following patients one morning.

1. Mrs. B complained of some deafness and a sense of fullness in her outer ear. Examination revealed that the wax in her ear canal had hardened and formed a plug of ..... _____.

2. Mr. J, aged 72, complained of gradually increasing deafness although he had no symptoms of pain or other problems related to the ears. Examination revealed that his deafness was the type called nerve deafness. The cranial nerve that carried impulses related to hearing to the brain is called the auditory nerve or the .... _____.

3. Baby L was brought in by his mother because he awakened crying, and was holding the right side of his head. He had been suffering from a cold but now, he seemed to be in pain. Examination revealed a bulging red eardrum. The eardrum is also called the .... _____.

CHAPTER NINE

The Heart

I. OVERVIEW

The ceaseless beat of the heart day and night throughout one's entire lifetime is such an obvious key to the presence of life that it is no surprise that this organ has been the subject of wonderment and poetry. When the heart stops pumping, life ceases. The cells must have oxygen and it is the heart's pumping action which propels oxygenated blood to them.

In size the heart has been compared to a *closed fist*. In location it is thought of as being on the left side although about one third is to the right of the midline. The muscular apex of the triangular heart is definitely on the left. It rests *on the diaphragm*, the dome-shaped muscle that separates the chest (thoracic) cavity from the abdominal space.

The heart of birds and mammals including man has 2 sides in which the *aerated* (so-called pure) blood and the *unaerated* (lower in oxygen) blood are kept *entirely separated* from each other. So the heart is really a *double pump* in which the 2 parts pump in unison, a vital duet. Each *side* of the heart is divided into 2 parts or *chambers* though here there is direct communication. The upper chamber in each case opens directly into the lower chamber, the ventricle. The 2 ventricles pump blood to organs of the body so their walls are much thicker than the walls of the upper chambers. The *coronary arteries* supply blood to the heart muscle itself.

II. TOPICS FOR REVIEW

1. the heart as a pump
2. structure of the heart wall
    a. endocardium
    b. myocardium
    c. pericardium
3. parts of the heart
    a. septum
    b. chambers and valves
4. the conduction system of the heart
    a. sinoatrial node; pacemaker
    b. atrioventricular node
    c. atrioventricular bundle
5. normal heart sounds

III. MATCHING EXERCISES

Matching only within each group, print the answer in the space provided.

Group A

arteries	mitral valve	interatrial septum
tricuspid valve	veins	aortic valve
endocardium	myocardium	interventricular septum
	pericardium	pulmonary valve

1. The membrane of which the heart valves are formed and which lines the interior of the heart is called .... _____

2. By far the thickest layer in the heart wall is the muscular one, the .... _____

3. The outermost layer of the heart and the lining of the pericardial sac is .... _____

4. A partition, the septum, separates the 2 sides of the heart. The thin-walled upper part of this septum is the .... _____

5. The larger part of the partition between the 2 sides of the heart is the .... _____

6. Between the 2 right chambers of the heart lies the right atrioventricular valve. It is also called the .... _____

7. The left atrioventricular valve is thicker and heavier than the right; it is made of 2 flaps or cusps. It is called the ... _____

8. Situated between the right ventricle and the pulmonary artery is the valve that prevents blood on its way to the lungs from returning to the right ventricle. This is the ... _____

9. The valve that prevents blood from returning after the left ventricle has emptied itself is the .... _____

10. Blood is pumped to the lungs and body tissues through.... _____

11. Oxygenated blood from the lungs and deoxygenated blood from the body tissues is carried through the .... _____

Group B

automaticity	sinoatrial node	bundle of His
atria	lubb, dupp	systole
atrioventricular node	venous	diastole

104

1. The active phase of cardiac contraction is called .... ———————————.

2. Heart muscle is capable of contracting independently of nervous control. This property is called .... ———————————.

3. The brief resting period that follows the contraction phase of the heart cycle is .... ———————————.

4. Impulses in the heart follow a definite sequence, beginning in the pacemaker. The pacemaker is located in the upper right atrial wall and is called the .... ———————————.

5. Next the excitation wave travels throughout the muscles of the upper heart chambers causing them to contract. These are the .... ———————————.

6. Following this the second mass of conduction tissue (located in the septum) is stimulated. This is the .... ———————————.

7. Finally the ventricular musculature contracts in response to stimulation by the branching part of the conduction system which is the .... ———————————.

8. Numerous disorders may cause deviations, or changes, in normal heart sounds. These sounds may be described by the syllables .... ———————————.

IV. LABELING

For each of the following illustrations, print the name or names of each labeled part on the numbered lines.

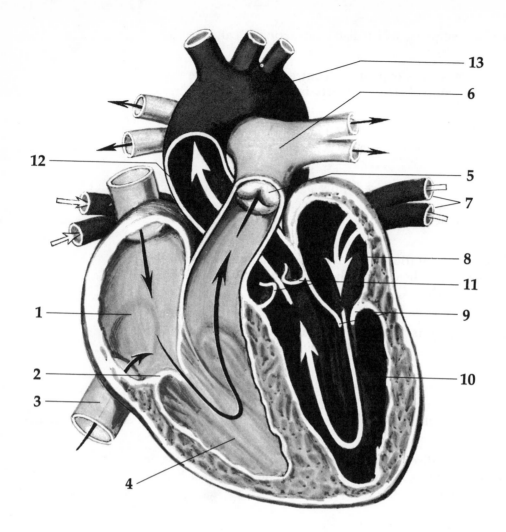

The heart and great vessels.

1. _____   8. _____

2. _____   9. _____

3. _____   10. _____

4. _____   11. _____

5. _____   12. _____

6. _____   13. _____

7. _____

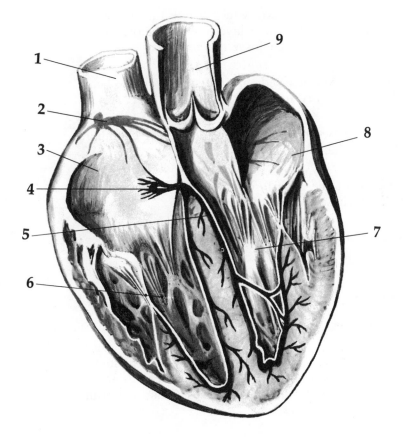

Conduction system of the heart.

1. _____   6. _____

2. _____   7. _____

3. _____   8. _____

4. _____   9. _____

5. _____

V. COMPLETION EXERCISE

Print the word or phrase that correctly completes the sentence.

1. The continuous one-way movement of the blood is known as the .... _____.

2. Each minute the heart contracts on an average of about _____.

3. Recall that one layer of serous membrane forms the lining of the closed sac; the other layer covers the organ surface. Lining the sac that encloses the heart is .... _____.

4. From Chapter 2 you learned that the serosa on the heart surface has the 2-word name .... _____.

5. The partition between the 2 thick-walled lower chambers of the heart is the .... _____.

6. Because each of the 3 parts of the 2 exit valves is half-moon shaped these valves are described as .... _____.

7. The left atrioventricular valve is called the .... _____.

VI. PRACTICAL APPLICATIONS

Study each discussion. Then print the appropriate word or phrase in the space provided.

1. Mrs. K had suffered from several attacks of rheumatic fever. The examining physician found evidence of damage to the membranous tissue of the valves. This membrane is called .... _____.

2. Using a stethoscope the physician listened for the heart sounds. The first sound occurs during the contraction phase of the cardiac cycle. This active phase is called .... _____.

3. The second sound includes the effects of valve closure and occurs during the resting part of the cycle. This phase is called .... _____.

4. Mr. L, age 62, complained of weakness and fatigue. Tests revealed damage to the conduction system of the heart. Normally the beginning of the heartbeat occurs in the natural pacemaker, which is also called the .... _____.

5. The greatest damage was found in the bundle of His which is also known as the .... _____.

CHAPTER TEN

Blood Vessels and Blood Circulation

I. OVERVIEW

The blood vessels are classified, according to function, as *arteries*, *veins* or *capillaries*; the arteries and veins are subdivided into *pulmonary* vessels and systemic vessels.

The 2 arterial systems—the systemic and the pulmonary—can be likened to trees; each has a trunk, the aorta for one and the pulmonary artery for the other. Each trunk has subdivisions, large and small branches that carry the blood into the capillaries where exchanges between the blood and the tissue fluid occur. The tissue fluid provides for the transfer of substances needed by the cell in exchange for those not to be used or those manufactured for use elsewhere in the body. The venous systems consist of small and finally larger tributaries that return the blood to the heart, which pumps it into the arterial trunks.

The *pulse* rate and the *blood pressure* are manifestations of the circulation; they tell the trained person a great deal about the overall condition of the individual being examined.

II. TOPICS FOR REVIEW

1. arteries: structure and function
2. veins: structure and function
3. capillaries: structure and function
4. pulmonary vessels
5. systemic vessels
6. branches of aorta
   a. ascending
   b. aortic arch
   c. thoracic
   d. abdominal
   e. iliac
   f. other parts of arterial tree
7. anastomoses

8. systemic veins
   a. superficial
   b. deep
   c. superior vena cava
   d. sinuses
   e. inferior vena cava
9. portal circulation
10. capillaries
11. pulse
12. blood pressure

III. MATCHING EXERCISES

Matching only within each group, print the answer in the space provided. One answer may be used more than once and there may be some words that are not used at all.

Group A

systemic	endothelium	pulmonary
arteries	celiac	carotid
aorta	capillaries	coronary

1. The vessels that are related to the lungs, including the arteries and their branches in the lungs and the veins that drain lung capillaries are all designated as . . . . _____.

2. Exchanges between the blood and the cells take place through the . . . . _____.

3. Since their function is to carry blood from the heart's pumping chambers, the thickest walls are those of the blood vessels called . . . . _____.

4. The innermost tunic of the artery is composed of . . . . _____.

5. The largest artery in the body is divided into 4 regions. This vessel is the . . . . _____.

6. The arteries that carry food and oxygen to the tissues of the body are classified as . . . . _____.

7. The ascending aorta has 2 branches that supply the heart muscle. Because they form a crown around the base of the heart they are classified as . . . . _____.

8. Supplying the head and neck on each side is an artery named the . . . . _____.

9. One of the unpaired arteries that supplies some of the viscera of the upper abdomen is a short trunk, the . . . . . . _____.

Group B

phrenic artery
brachiocephalic trunk
anastomosis
renal arteries

brachial artery
lumbar arteries
right subclavian artery
hepatic artery

common iliac arteries
left common carotid artery
superior mesenteric artery

1. Coming off the aortic arch is a short artery formerly called the innominate artery. This is the .... —————————.

2. Supplying the left side of the head and neck is the ...... —————————.

3. Oxygenated blood is carried to the liver by the .... —————————.

4. The largest branch of the abdominal aorta supplies most of the small intestine and the first half of the large intestine. This branch is the .... —————————.

5. The muscular partition between the abdominal and thoracic cavities is supplied by a right and a left .... —————————.

6. The artery supplying the arm is a continuation of the axillary artery, and is called the .... —————————.

7. Blood supply to the right upper extremity is through the —————————.

8. The largest of the paired branches of the abdominal aorta are those that supply the kidneys. These are the .... —————————.

9. The abdominal aorta finally divides into 2 .... —————————.

10. A communication between 2 arteries is called an .... —————————.

11. Supply to the abdominal wall is through the .... —————————.

Group C

brachiocephalic trunk
mesenteric
paired
celiac trunk

radial artery
femoral artery
circle of Willis

unpaired
basilar artery
volar arch

1. An anastomosis of the 2 internal carotid arteries and the basilar artery is located immediately under the center of the brain. It is called the .... —————————.

2. The inferior mesenteric is an example of an artery that is —————————.

111

3. The radial and ulnar arteries in the hand anastomose to form the .... _____.

4. Anastomoses between branches of the vessels supplying blood to the intestinal tract comprise arches named ..... _____.

5. The right subclavian artery and the right common carotid artery are branches of the .... _____.

6. The left gastric artery and the splenic artery are 2 of the 3 branches of the .... _____.

7. The union of the 2 vertebral arteries forms the .... _____.

8. The external iliac arteries extend into the thigh. Here each of them becomes a .... _____.

9. The popliteal arteries are examples of the many blood vessels that are .... _____.

10. The branch of the brachial artery that extends down the forearm and wrist of the thumbs side is the .... _____.

Group D

azygos vein	median cubital	saphenous vein
inferior vena cava	portal vein	superior vena cava
jugular veins	brachiocephalic veins	venous sinuses
liver		

1. The longest vein is the superficial one called the .... _____.

2. Because of its location near the surface at the front of the elbow one of the veins frequently used for removing blood for testing is the .... _____.

3. The areas supplied by the carotid arteries are drained by the .... _____.

4. The union of the jugular and subclavian veins forms the _____.

5. Veins draining the head, the neck, the upper extremities and the chest all empty into the .... _____.

6. Before reaching the superior vena cava (and then the heart), blood from the chest wall drains into the .... _____.

7. Structures other than veins also drain deoxygenated blood. These structures are called .... _____.

8. The blood from the parts of the body below the diaphragm is drained by the large vein called the .... _____.

9. Tributaries from the unpaired organs empty into a vein that enters the liver where it subdivides into smaller veins. This unusual vein is called the .... _____.

10. Food products are released into the circulatory system from the .... _____.

Group E

sinusoids
coronary sinus
common iliac veins
portal tube
superior mesenteric vein

hepatic veins
lateral sinuses
superior sagittal sinus
left testicular vein

capillary walls
cavernous sinuses
gastric veins
veins

1. The inferior vena cava begins with the union of the 2 .... _____.

2. The only exceptions to the rule that paired veins empty directly into the vena cava are the left ovarian vein and the .... _____.

3. Unpaired veins coming mostly from the digestive tract are drained by a special vein called the .... _____.

4. Among the paired veins that empty directly into the inferior vena cava are those draining the liver, the ...... _____.

5. The vein that drains most of the small intestine and the first part of the large intestine is the .... _____.

6. The tributaries of the portal tube include those that drain the stomach, the .... _____.

7. Within the liver, there are no capillaries; instead, this function is performed by .... _____.

8. Cell nutrients pass into the tissue fluid through the ..... _____.

9. The veins of the heart wall drain mainly into the .... _____.

10. The ophthalmic veins drain into the .... _____.

11. Nearly all the blood from the veins of the brain eventually empties into one or the other of the transverse or .... _____.

12. Metabolic waste products proceed through capillary walls into the blood of the capillary and then into the .... _____.

13. In the midline above the brain and in the fissure between the 2 cerebral hemispheres is a long blood-containing space called the .... _____.

Group F

faster   dorsalis pedis   radial artery
hypotension   pulse   sphygmomanometer

1. Beginning at the heart and traveling along the arteries is a wave of increased pressure started by the force of ventricular contractions. This wave is called the .... _____.

2. The wave is readily felt at the wrist because of the artery that passes over the bone on the thumb side. This is the.. _____.

3. Sometimes it is necessary to use the artery on the top of the foot for obtaining the pulse. This is the .... _____.

4. Blood pressure is recorded by the .... _____.

5. It is important to recognize factors that may influence pulse rate. Emotional disturbance, for example, may cause the pulse rate to be .... _____.

6. An abnormal decrease in blood pressure, as may occur in shock, is called .... _____.

IV. LABELING

For each of the following illustrations, print the name or names of each labeled part on the numbered lines.

1. _____ 11. _____

2. _____ 12. _____

3. _____ 13. _____

4. _____ 14. _____

5. _____ 15. _____

6. _____ 16. _____

7. _____ 17. _____

8. _____ 18. _____

9. _____ 19. _____

10. _____ 20. _____

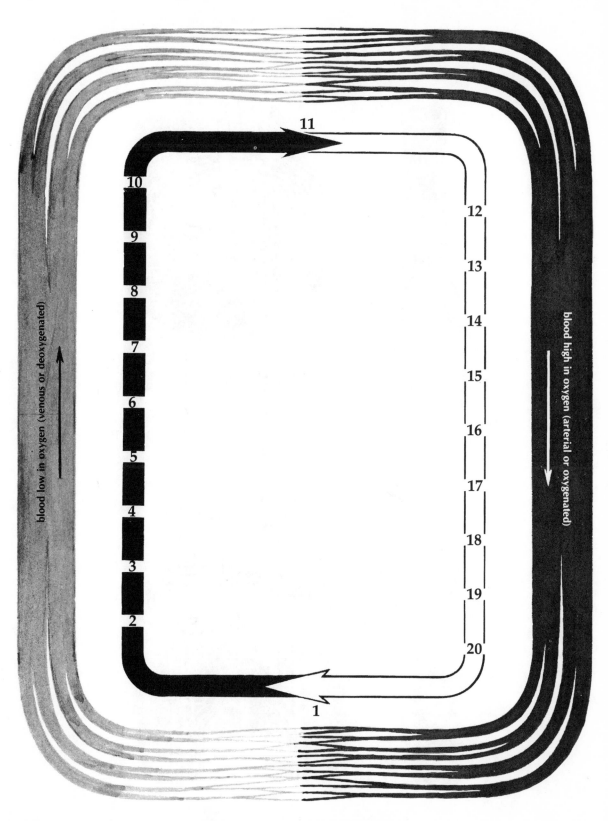

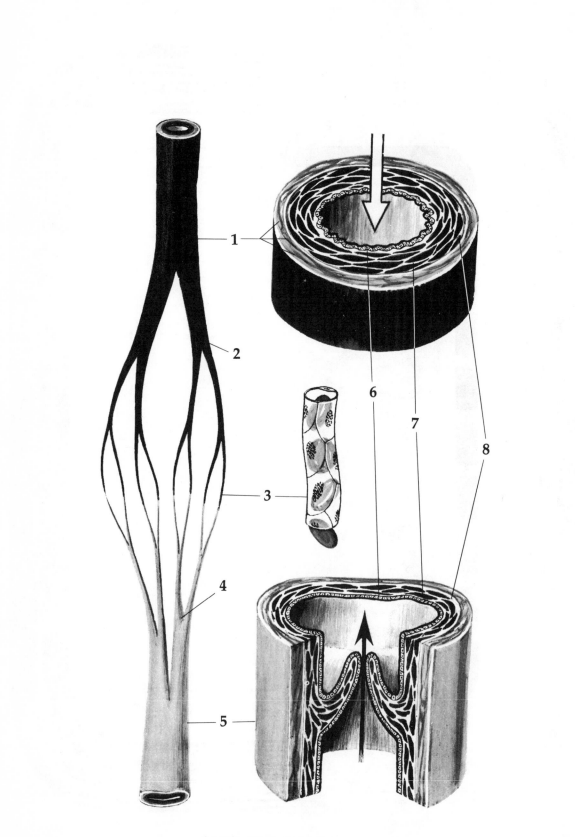

Sections of small blood vessels.

1. _____   5. _____

2. _____   6. _____

3. _____   7. _____

4. _____   8. _____

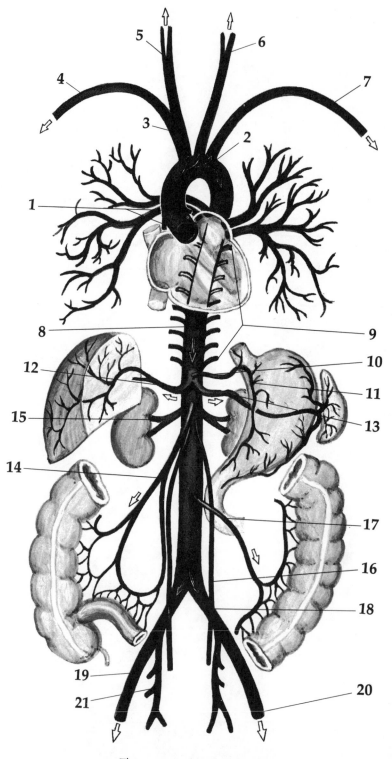

The aorta and its branches.

1. _____

2. _____

3. _____

4. _____

5. _____

6. _____

7. _____

8. _____

9. _____

10. _____

11. _____

12. _____

13. _____

14. _____

15. _____

16. _____

17. _____

18. _____

19. _____

20. _____

21. _____

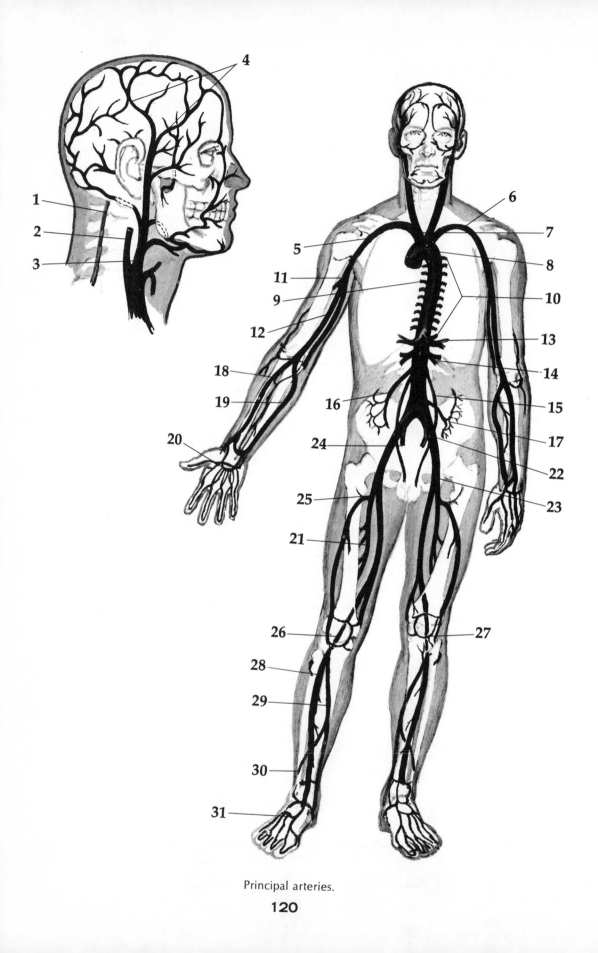

Principal arteries.

1. _____

2. _____

3. _____

4. _____

5. _____

6. _____

7. _____

8. _____

9. _____

10. _____

11. _____

12. _____

13. _____

14. _____

15. _____

16. _____

17. _____

18. _____

19. _____

20. _____

21. _____

22. _____

23. _____

24. _____

25. _____

26. _____

27. _____

28. _____

29. _____

30. _____

31. _____

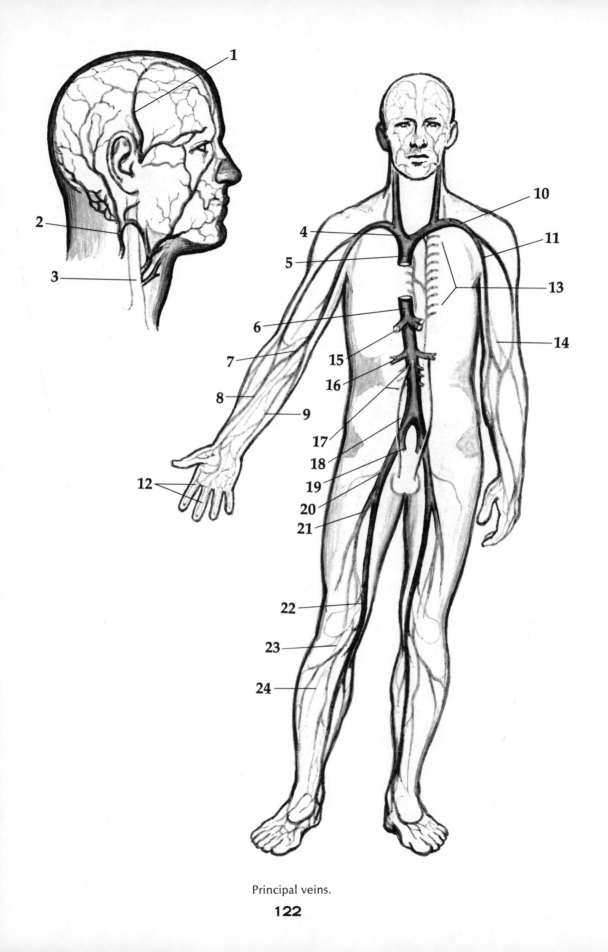

Principal veins.

1. _____

2. _____

3. _____

4. _____

5. _____

6. _____

7. _____

8. _____

9. _____

10. _____

11. _____

12. _____

13. _____

14. _____

15. _____

16. _____

17. _____

18. _____

19. _____

20. _____

21. _____

22. _____

23. _____

24. _____

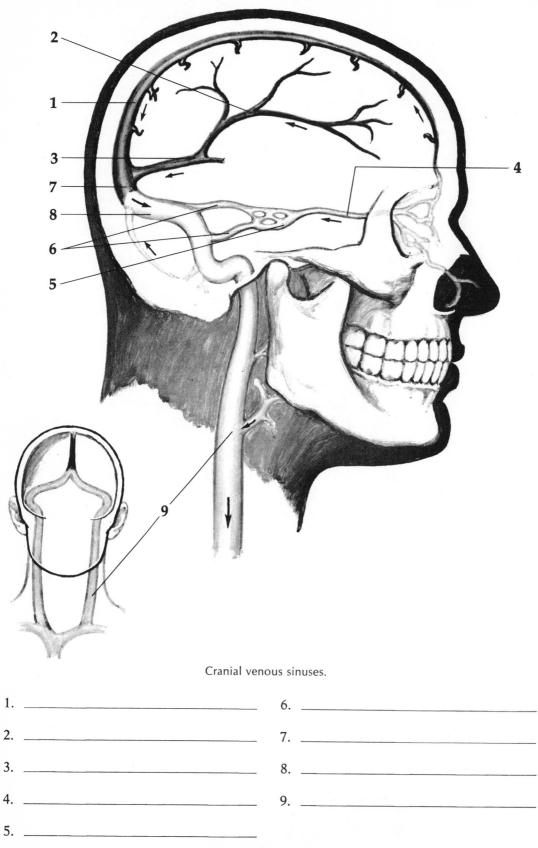

Cranial venous sinuses.

1. _____   6. _____

2. _____   7. _____

3. _____   8. _____

4. _____   9. _____

5. _____

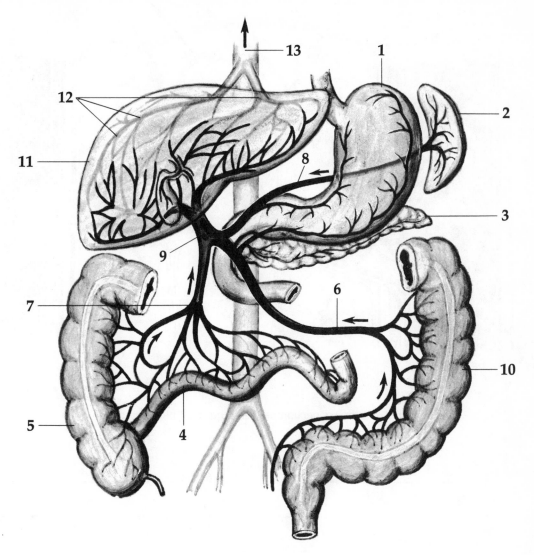

Portal circulation.

1. _____   8. _____

2. _____   9. _____

3. _____   10. _____

4. _____   11. _____

5. _____   12. _____

6. _____   13. _____

7. _____

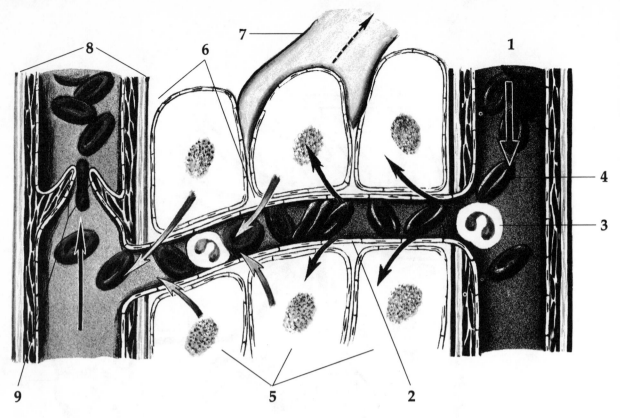

Diagram showing the connection between the small blood vessels through capillaries.

1. _____   6. _____

2. _____   7. _____

3. _____   8. _____

4. _____   9. _____

5. _____

V. COMPLETION EXERCISE

Print the word or phrase that correctly completes the sentence.

1. Deoxygenated blood is carried from the right ventricle by the .... _____.

2. The smallest subdivisions of arteries have thin walls in which there is little connective tissue and relatively more muscle. These vessels are .... _____.

126

3. To supply nutrients to body tissues and carry off waste products from the tissues are functions of the circulation described as .... ——————————.

4. The middle tunic of the arterial wall is composed of elastic connective tissue plus .... ——————————.

5. Endothelium comprises the tunic of the arterial wall that is .... ——————————.

6. The volar arch in the hand is formed by the union of the radial artery with the .... ——————————.

7. Communications between the branches of vessels that supply blood to the intestinal tract are called the .... ——————————.

8. The longest veins in the body are the superficial ones in the lower extremities. They are the .... ——————————.

9. The smallest veins are formed by the union of capillaries. These tiny vessels are called .... ——————————.

10. The circle of Willis is formed by a union of the internal carotid arteries and the basilar artery. Such a union of end arteries is called an .... ——————————.

VI. PRACTICAL APPLICATIONS

Study each discussion. Then print the appropriate word or phrase in the space provided.

1. Mr. S, age 53, complained of shortness of breath, weakness and pain in the left chest. Examination indicated that the left semilunar valve was not functioning properly. This valve guards the entrance into the largest artery which is the .... ——————————.

2. Mr. K, age 71, was admitted to the hospital because of fainting attacks and inability to recall events just before and after these episodes. Physical examination revealed some interference with the normal flow through the large arteries that carry blood to the brain. These neck arteries are named .... ——————————.

3. Small amounts of blood could still reach Mr. K's brain through the 2 vertebral arteries. These join at the base of the brain to form a single artery called the .... ——————————.

4. Mr. K was protected from brain death by the fact that 3 arteries join to form an anastomosis under the brain. This anastomosis is the .... ——————————.

5. Mr. J, an alcoholic, came in complaining of a variety of digestive and nervous disturbances. Examination revealed an enlarged liver and an accumulation of fluid in the abdominal cavity. There was evidence of obstruction of the large vein that drains the unpaired organs of the abdomen. This large vessel is called the .... _____.

6. Mr. J and his digestive problems were studied for evidence of back pressure within the veins that drain the intestine. The larger of these is called the .... _____.

7. Mrs. B, age 67, had been diabetic for several years. Now there was evidence of artery disease in several of the larger vessels. One of these is named according to regions including the portion nearest the heart, called the .... _____.

8. Another vessel that was seriously damaged by hardening was the short one that is the first branch of the aortic arch. It is the .... _____.

9. Mrs. B had also noted difficulty in walking. An investigation revealed disease of the paired branches of the terminal abdominal aorta, the .... _____.

10. Mr. M had spent many years working in a store as a salesman. He had varicose veins involving the longest superficial vein in the lower extremity. This is known as the .... _____.

11. Following the injection of these surface veins, Mr. M depended on deep veins for the return of blood from the extremities. Among these were the deep veins of the thigh called the .... _____.

CHAPTER ELEVEN

The Lymphatic System and Lymphoid Tissue

I. OVERVIEW

Lymph is the watery fluid that flows within the lymphatic system. It originates from the blood plasma and from the tissue fluid that is found in the minute spaces around and between the body cells. Lymph may contain certain cellular waste products as well as fat globules from the digestive system following a meal. The relatively few cells that are present are usually lymphocytes. The fluid moves from the *lymphatic capillaries* through the *lymphatic vessels* and thence to the *right lymphatic duct* and the *thoracic duct*. The lymphatic vessels are thin-walled and delicate; like some veins, they have valves that prevent backflow of lymph.

The *lymph nodes*, which are the system's filters, are composed of *lymphoid tissue*. These nodes remove small foreign bodies such as dead blood cells, carbon particles and pathogenic organisms; they also manufacture *lymphocytes* and *antibodies*. Chief among them are the *cervical nodes* in the neck, the *axillary nodes* in the armpit, the *tracheobronchial nodes* near the trachea and bronchial tubes, the *mesenteric nodes* between the peritoneal layers, and the *inguinal nodes* in the groin area.

In addition to these are several organs of lymphoid tissue whose functions are somewhat different. The *tonsils* filter tissue fluid. The *thymus* is essential for antibody formation and development of immunity during the early weeks of life. The *spleen* has numerous functions, among which are the destruction of used-up blood cells, as a reservoir for blood, and the production of red cells before birth.

A distinctive attribute of lymphoid tissue is its abundance. Many more of these masses are normally present than are actually needed, so that removal of certain of them does not interfere with the overall functioning of the human organism.

II. TOPICS FOR REVIEW

1. lymphatic system
   a. lymph

              (1) composition
              (2) function
        b. lymph conduction
    2. lymphoid tissue
        a. location
        b. functions
        c. 5 main groups of lymph nodes
        d. other lymphoid structures differing in function from main groups

III. MATCHING EXERCISES

Matching only within each group, print the answer in the space provided.

Group A

right lymphatic duct	blood	inguinal nodes
lacteals	valves	chyle
endothelium	buboes	axillary nodes
cervical nodes		spleen

1. There is easy passage of soluble materials and water
    through the walls of lymphatic capillaries, in which a
    single layer of cells forms the.......................... _____.

2. The lymphatics resemble some veins in that they contain
    structures that prevent backflow. These are............. _____.

3. One pathway for fats from digested food to the blood-
    stream is through specialized lymphatic capillaries of the
    intestine that are called................................ _____.

4. Lymph is drained from the right side of the head, of the
    neck, of the thorax and of the right upper extremity by
    the .... _____.

5. The combination of fat globules and lymph gives rise to
    a milky-appearing fluid called........................... _____.

6. Lymph nodes are named according to location. Those
    located in the armpits are known as..................... _____.

7. Drainage of lymph from the lower extremities and the
    external genitalia is through the....................... _____.

8. Abnormally large inguinal nodes, as may be found in
    certain infections, are called........................... _____.

9. The final destination of filtered lymph is the............ _____.

10. The lymph nodes located in the neck and draining certain
    parts of the head and neck are known as................ _____.

130

11. During embryonic and fetal life red blood cells are produced in the .... _____.

Group B

thymus	pharyngeal tonsils	bile
spleen	lingual tonsils	lymph
	palatine tonsils	

1. The oval lymphoid bodies located at each side of the soft palate are known as .... _____.

2. The enlarged masses of lymphoid tissue often found on the back wall of the pharynx and commonly called adenoids are correctly called.......................... _____.

3. At the back of the tongue are masses of lymphoid tissue called .... _____.

4. The structure that is believed to be essential in the formation of antibodies very early in life is the.............. _____.

5. Blood filtration is carried out by an organ located in the upper left quadrant (left hypochondriac region) of the abdomen. This is the................................ _____.

6. The end products of red blood cell destruction are returned to the liver, where they are manufactured into a secretion called .... _____.

7. The watery fluid that flows within the lymphatic system is called .... _____.

Group C

lymph nodes	backflow	radial lymphatic vessels
phagocytosis	drainage	lymph
subclavian vein	antibodies	macrophages
thoracic duct		

1. Located in the bone marrow, the spleen and lymph nodes are the cells that absorb and destroy foreign matter. These are .... _____.

2. An important function of lymphoid tissue is the production of chemical substances that aid in combatting infection. These are called............................ _____.

3. The spleen generates cells that are able to engulf bacteria and other foreign cells. This process is known as........ _____.

131

4. The fluid that moves from blood plasma to the tissue spaces and finally to special collection vessels is called... _____.

5. Before the lymph reaches the veins it is passed through organs that act as filters. These are.................... _____.

6. The lymphatic vessels serve as a system for............. _____.

7. Lymph received in the right lymphatic duct drains into the right .... _____.

8. Lymph is drained from the body below the diaphragm and on the left side above the diaphragm by the largest lymphatic vessel, the .... _____.

9. The valves of the lymphatic vessels prevent lymph....... _____.

10. Lymphatic vessels are named according to location; thus, those on the lateral side of the forearm are the.......... _____.

Group D

lymphocytes	hilum	plasma
lymph	lacteals	lymph nodes
veins	blindly	cisterna chyli

1. Lymphatic capillaries differ from blood capillaries in that they begin .... _____.

2. The first part of the thoracic duct is enlarged, forming a temporary storage area. It is called the................ _____.

3. Chyle, the fluid formed by combination of lymph and fat globules, comes from the intestinal..................... _____.

4. An important function of lymph nodes is the manufacture of white blood cells known as......................... _____.

5. Intercellular fluid originates from the liquid part of the blood. The liquid part of the blood is called............. _____.

6. Tissue fluid passes from the intercellular spaces into the lymphatic vessels; it is then called..................... _____.

7. The masses of lymphoid tissue that filter foreign substances from the liquid lymph are known as .... _____.

8. Both superficial and deep vessels are found in the lymphatic system just as in the case of the system of........ _____.

9. The area of exit for the vessels carrying lymph out of the node is known as the . . . . ———————————————.

IV. LABELING

Print the name or names of each labeled part on the numbered lines.

1. _____

2. _____

3. _____

4. _____

5. _____

6. _____

7. _____

14. _____

15. _____

16. _____

17. _____

8. _____

9. _____

10. _____

11. _____

12. _____

13. _____

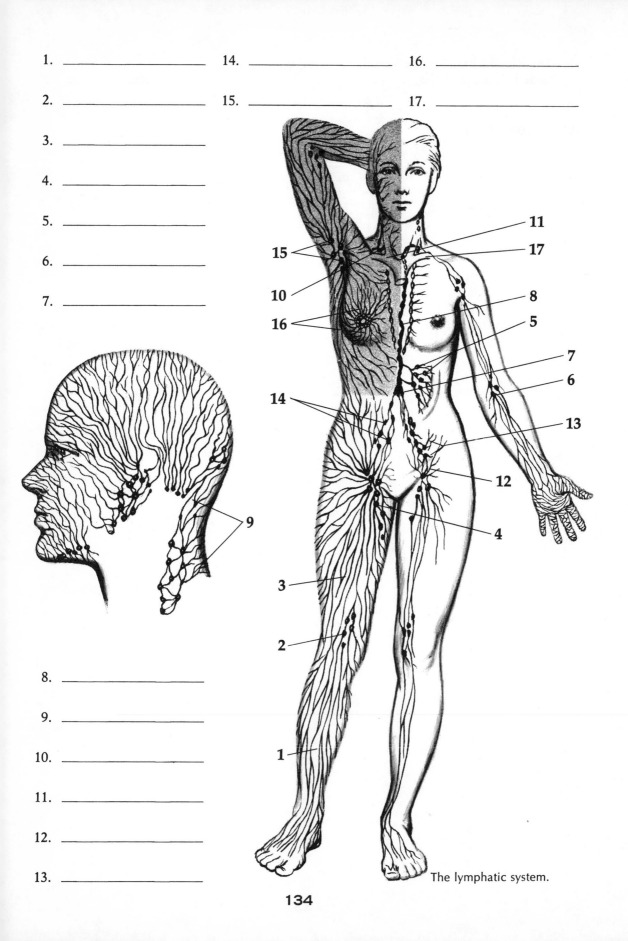

The lymphatic system.

V. COMPLETION EXERCISE

Print the word or phrase that correctly completes the sentence.

1. All parts of the body except those above the diaphragm on the right side are drained by the largest lymph channel called the .... _____.

2. Lymphatic vessels from the left side of the head, neck, and thorax empty into the largest of the lymph vessels, the _____.

3. The lymph from the body below the diaphragm and from the left side above the diaphragm is carried into the blood of the .... _____.

4. Between the 2 layers of peritoneum that form the mesentery are found nodes called.......................... _____.

5. In city dwellers nodes may appear black because they become filled with carbon particles. This is true mostly of the nodes that surround the windpipe and its divisions. These are the .... _____.

6. A disease prevalent during the Middle Ages was responsible for the death of thousands of people. This disease was characterized by the presence of buboes—inflammatory swellings of the inguinal nodes—so it was called.... _____.

7. The structure popularly known as adenoids is correctly called the .... _____.

8. The spleen and other organs produce cells that can engulf harmful bacteria and other foreign cells, by a process called .... _____.

9. The enlargement at the beginning of the thoracic duct is called the .... _____.

10. Lymph and fat enter the bloodstream at the end of the thoracic duct as it empties into the vein, the........... _____.

VI. PRACTICAL APPLICATIONS

1. Mrs. B, age 38, underwent biopsy of a small mass in her right breast which was positive for cancer. She was now being admitted in order to have a radical mastectomy. In this operation certain nodes are removed as well as the breast, because cancer cells from the breast often invade them. These are the armpit nodes called the............. _____.

2. Mr. G, age 31, complained of swellings in his neck, his armpits, his groin and other areas. A diagnosis of Hodgkin's disease was made. Treatment with radiation and antineoplastic drugs was planned. The nodes of the neck are designated the .... _____.

3. Mr. K, age 41, had been hunting wild rabbits in the central valley of California. Several days after dressing a number of these rabbits an ulcer developed on his hand. A tentative diagnosis of tularemia, or rabbit fever, was made. The infecting organisms had been carried to the axillary nodes via colorless tubes called............................ _____.

4. Mr. J, age 21, was suffering from a venereal disease. The lymph nodes located in the groin region were enlarged. The regional name for these nodes is.................... _____.

5. Mrs. G brought her 2-month-old infant to the clinic because he was so susceptible to infections. One important function of lymph nodes is to aid in combatting infections by the production of chemical substances known as...... _____.

6. Mrs. G's infant was examined for the size of an organ located in the lower neck and upper anterior chest region. It is an important structure related to the development of immunity in fetal life and in early infancy. This organ is called the .... _____.

7. Studies of Mrs. G's infant included blood counts. The numbers of white cells were noted and particularly those produced by lymphoid tissue, namely the.............. _____.

CHAPTER TWELVE

The Digestive System

I. OVERVIEW

The complex process by which the food we eat reaches the cells throughout the body is accomplished through *digestion* and *absorption*. These are functions of the *digestive system*; its components are the *alimentary canal* and the *accessory organs*.

The alimentary canal, consisting of the *mouth*, the *pharynx*, the *esophagus*, the *stomach* and the large and small *intestine*, forms a continuous passageway in which ingested food is prepared for utilization by the body, and waste products are collected to be expelled from the body. The *liver*, the *gallbladder*, and the *pancreas*, which comprise the accessory organs, manufacture various substances needed to regulate food metabolism, serve as storage areas for certain substances which are released as needed, and function in other ways to help maintain a normal state of health.

Since ingested food is the main source of nourishment for the body, a *balanced diet*, with avoidance of "fads in foods," should be followed by all persons who are in a normal state of health.

II. TOPICS FOR REVIEW

1. components of the digestive system
2. oral cavity
3. peristalsis
4. swallowing tubes and accessories
5. stomach
6. small intestine
7. large intestine
8. liver
9. gallbladder
10. pancreas
11. peritoneum
12. nutrition and diet

III. MATCHING EXERCISES

Matching only within each group, print the answer in the space provided. Some answers may be used more than once.

Group A

tongue   permanent teeth   deciduous teeth
incisors   alimentary canal   molars
absorption   digestion

tongue   permanent teeth   deciduous teeth
incisors   alimentary canal   molars
absorption   digestion

1. The process by which ingested food is converted into substances that may be taken into the cells is called........ _____.

2. The transfer of digested food to the bloodstream is called _____.

3. The structures and organs through which ingested food or its breakdown products pass comprise the.............. _____.

4. The mouth, the pharynx, the esophagus, the stomach and the intestines are part of the.......................... _____.

5. One can differentiate taste sensations by means of special organs of the .... _____.

6. The grinding teeth located in the back part of the oral cavity are called .... _____.

7. The temporary baby teeth are lost, and are therefore described as .... _____.

8. During the time that the baby teeth are appearing, or erupting, the second set of teeth are developing in the jawbones. These are the.............................. _____.

9. The first 8 of the baby teeth to appear are the............. _____.

10. The wisdom teeth usually appear during the later teen years. They are more accurately described as the third... _____.

11. The cutting teeth located in the front portion of the buccal cavity are the.............................. _____.

Group B

first molars   third molars   32 teeth
second molars   premolars   20 teeth
oral cavity   canines

1. The baby molar teeth are not replaced by permanent molars but by smaller................................... _____.

2. Decay and infection of baby molar teeth may easily spread to the first permanent teeth, or the.............. _____.

3. The so-called eyeteeth are the......................... _____.

4. The incisors are located in the front part of the.......... _____.

5. Normally, at about age 12 the jawbones should be large enough to accommodate the erupting................... _____.

6. It sometimes happens that the jaw is not large enough to accommodate the last teeth to erupt. These teeth are identified as the .... _____.

7. By the time the baby is 2 years old he should have all the deciduous teeth. This means there are................ _____.

8. The molar teeth are found in the back portion of the..... _____.

9. An adult who has a full set of permanent teeth has....... _____.

Group C

saliva	pharynx	alveolus
esophagus	pyorrhea alveolaris	mastication
gingivitis	stomatitis	uvula
caries	deglutition	peristalsis

1. Infection of the gum is known as...................... _____.

2. Infection of the mucous membrane lining of the mouth, except for the gums, is called......................... _____.

3. Certain inflammations involve the tooth pocket, or....... _____.

4. An inflammation of the tooth socket which is associated with a discharge of pus is called...................... _____.

5. Loss of teeth is often the result of tooth decay, or dental _____.

6. The process of chewing is called...................... _____.

7. The act of swallowing is known as.................... _____.

8. An essential part of the digestive process involves the coating of the food with mucus and the dissolving of the food in the mouth by the digestive juice............... _____.

9. Food and liquid are prevented from entering the nasal cavities by the soft fleshy........................... _____.

10. Food is propelled along the alimentary canal by the rhythmic motion known as........................... _____.

11. The tongue pushes the food into the.................... _____.

12. The contents of the alimentary canal are moved along as far as the stomach by the gullet, or.................... _____.

Group D

cardiac valve   pharynx   soft palate
pyloric sphincter   chyme   epiglottis
                      gastric juice   rugae

1. The uvula hangs from the back of the roof of the oral cavity. This part of the oral cavity roof is the........... _____.

2. During deglutition there is contraction of the muscles of the .... _____.

3. With the muscular contraction that occurs during deglutition, the openings into the air spaces above and below the mouth are closed off by the soft palate and by the....... _____.

4. The structure that guards the entrance into the stomach is called the .... _____.

5. The valve between the distal end of the stomach and the small intestine is the................................ _____.

6. If the stomach is empty, there will be many folds in the lining. These folds are called........................ _____.

7. Contents in the stomach are mixed with hydrochloric acid and enzymes to form.................................. _____.

8. The combination of hydrochloric acid and enzymes in the stomach is referred to as........................... _____.

Group E

bicuspids   villi   ileocecal
hypoacidity   hepar   sugars
molars   ileum   fats
jejunum   hyperacidity   duodenum

1. An abnormally high production of stomach acid is called _____.

2. The permanent premolars are also called................ _____.

3. The first part of the small intestine is the.............. _____.

4. The premolar teeth replace the deciduous.............. _____.

5. Storage of simple sugar in the form of glycogen is one function of the .... _____.

6. The absorbing area of the small intestinal mucosa is greatly increased by numerous projections called........ _____.

7. Lying just beyond the duodenum is the second part of the small intestine, the .... _____.

8. Although bile contains no enzymes, it aids in the digestion of .... _____.

9. The final, and longest, section of the small intestine is the _____.

10. Because of its location the valve between the small and large intestine is described as......................... _____.

11. An abnormally low production of stomach acid may be an indication that serious illness is present. Such underacidity is called .... _____.

12. Among the classes of nutrients that are essential to cell life are carbohydrates, which include................. _____.

Group F

glycogen	hepar	hepatitis
heparin	trypsin	lipase
fibrinogen	mesentery	amylopsin

1. Nerves, arteries and other structures supplying the small intestine are found between the 2 layers of peritoneum called the .... _____.

2. The largest gland in the body is the liver, or............ _____.

3. Inflammation of the liver is called.................... _____.

4. The liver has many essential functions. One of these is the manufacture of a substance which prevents clotting of the blood. This is................................ _____.

5. Sugar is stored by the liver and released as simple sugar (glucose) as needed. The form in which sugar is stored is _____.

6. The liver has many functions including the production of plasma proteins such as albumin and.................. _____.

7. Pancreatic juice contains enzymes that act in various ways on the chyme in the small intestine. Starch is changed to sugar by the pancreatic enzyme .... _____.

8. Fats must be broken down into simpler compounds in order to be readily absorbed. The enzyme responsible for this breaking down is................................ _____.

141

9. Proteins enter the bloodstream in the form of amino acids. The splitting of proteins is accomplished by the pancreatic enzyme .... _____.

Group G

liver   lacteals   fecal matter
ptyalin   ducts   small intestine
cecum   vermiform appendix   colon

1. The important function of absorption is carried out through the numerous villi projecting from the mucosa of the .... _____.

2. Following their absorption into the bloodstream through the capillary walls of the villi, food materials are stored and released as needed by the .... _____.

3. In the saliva there is an enzyme that begins starch digestion. It is called .... _____.

4. Much of fat absorption occurs through the lymphatic capillaries of the villi; they are called .... _____.

5. Digestive juices are carried from accessory organs of digestion to the duodenum by means of .... _____.

6. The materials to be eliminated will continue through the ileocecal valve into the beginning of the large intestine. Here it enters a small pouch called the .... _____.

7. The small blind tube attached to the proximal part of the large intestine is called the .... _____.

8. In the large intestine, layers of involuntary muscle move the solid waste products on toward the rectum. This waste material is called .... _____.

9. The longer part of the large intestine is the .... _____.

IV. LABELING

For each of the following illustrations, print the name or names of each labeled part on the numbered lines.

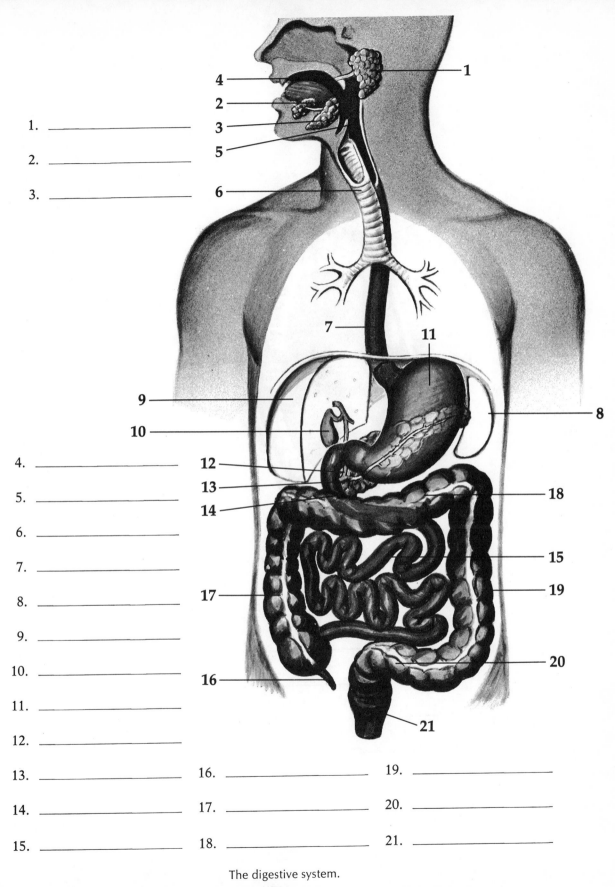

1. _____

2. _____

3. _____

4. _____

5. _____

6. _____

7. _____

8. _____

9. _____

10. _____

11. _____

12. _____

13. _____   16. _____   19. _____

14. _____   17. _____   20. _____

15. _____   18. _____   21. _____

The digestive system.

143

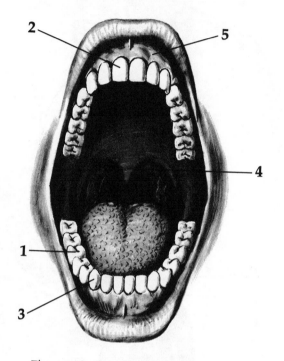

The mouth, showing teeth and tonsils.

1. _____

2. _____

3. _____

4. _____

5. _____

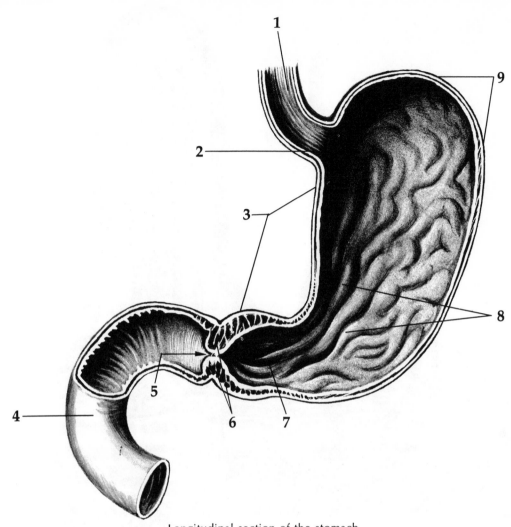

Longitudinal section of the stomach.

1. _____   6. _____

2. _____   7. _____

3. _____   8. _____

4. _____   9. _____

5. _____

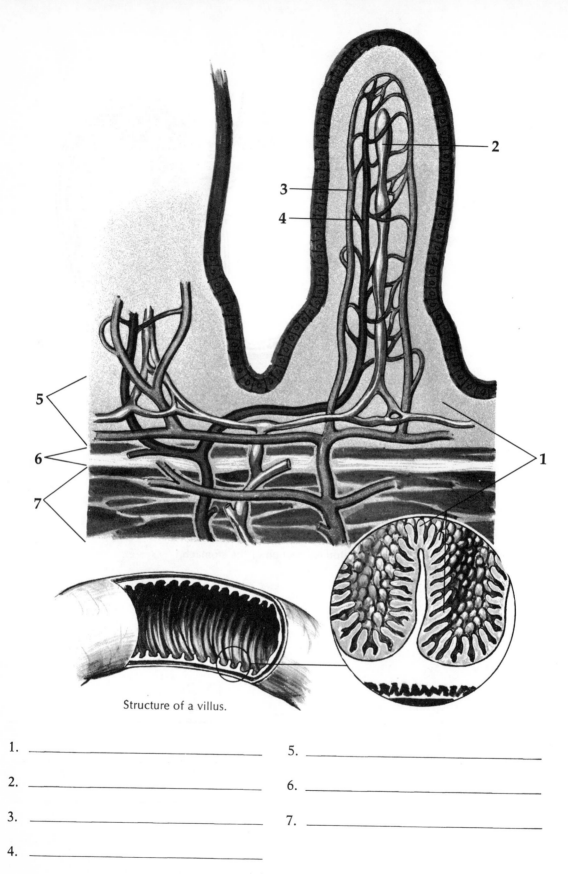

Structure of a villus.

1. _____   5. _____

2. _____   6. _____

3. _____   7. _____

4. _____

V. COMPLETION EXERCISE

Print the word or phrase that correctly completes the sentence.

1. The amount of sugar "burned" in the tissues is regulated by a pancreatic hormone called........................ ——————————.

2. Differentiation of bitter, salty, sweet and sour food sensations is accomplished through the .... ——————————.

3. Following digestion foods are transferred to the bloodstream by a process called........................... ——————————.

4. Discharge of pus from a tooth socket accompanied by inflammation is called .... ——————————.

5. Saliva is produced by 3 pairs of glands, of which the largest are the ones located near the angles of the jaw, called the .... ——————————.

6. The first permanent tooth to make its appearance in a child's mouth is the................................. ——————————.

7. The salivary glands located under the tongue are the..... ——————————.

8. One component of gastric juice kills bacteria and thus helps defend the body against disease. This substance is.. ——————————.

9. Most of the digestive juices contain substances that cause the chemical breakdown of foods without entering into the reaction themselves. These catalytic agents are.......... ——————————.

10. Starches and sugars are classified as.................... ——————————.

11. The highest concentration of fuel value is provided by... ——————————.

12. The exit valve from the stomach is the.................. ——————————.

13. The lower part of the colon bends into an S-shape so this part is called the.................................... ——————————.

14. A temporary storage section for indigestible and unabsorbable waste products of digestion is a tube called the .... ——————————.

15. The portion of the distal part of the large intestine called the anal canal leads to the outside through an opening called the .... ————————————.

16. The organ in which most digestion and practically all absorption of food occurs is the........................ ————————————.

17. The muscular sac in which bile is stored to be released as needed is called the............................. ————————————.

18. In the United States many problems related to nutrition result from use of the wrong foods. This condition is called .... ————————————.

VI. PRACTICAL APPLICATIONS

Study each discussion. Then print the appropriate word or phrase in the space provided.

1. Mr. C, age 36, complained of pain in the "pit of the stomach." Ingestion of food seemed to provide some relief. The physician ordered x-ray studies to be done. These studies indicated that the first part of the small intestine was involved. This short section is called the........... ————————————.

2. Mr. C was a tense man who felt it was important to do well in his business; he worked long hours. Because of this constant stress, his physician suspected that excessive acid was being produced. This important stomach acid is called .... ————————————.

3. Three-month-old John was brought to the clinic by his mother because he had suffered several bouts of vomiting and could not retain food. The tentative diagnosis was a constricted or spastic muscle of the exit valve, the....... ————————————.

4. Mr. K, age 24, complained of pain in his lower abdomen, in the right iliac region. Other symptoms and blood counts indicated that he was suffering from an inflammation of a wormlike appendage of the cecum. The 2-word name for this structure is............................ ————————————.

5. Mr. S, age 51, complained of blood in his fecal material. Investigation revealed a tumor in the last part of the colon. This S-shaped section of the colon is called the... ————————————.

6. Mrs. S, age 34, also complained of seeing blood at the time of defecation. Examination revealed enlarged veins called hemorrhoids located in the distal part of the large intestine immediately following the rectum. This leads to the outside through an opening called the............... ————————————.

7. Mr. G, age 28, had felt under par for some time. Studies showed that Mr. G had a virus infection of the liver, a disease called hepatitis. Among the many functions of the liver that can be affected in liver disorders is the production of a digestive juice called....................... ——————————————.

CHAPTER THIRTEEN

The Respiratory System

I. OVERVIEW

Oxygen is supplied to the tissue cells and *carbon dioxide* is removed from them through the arrangement of spaces and passageways known as the *respiratory system*. This system comprises the *nasal cavities*, the *pharynx*, the *larynx*, the *trachea* and the *lungs*.

The 2 phases of breathing are *inhalation*—the drawing in of air—and *exhalation*—the expulsion of air. The rate normally varies from 12 to 25 times per minute. Breathing is under the control of the *respiratory center* in the brain located in the *medulla*.

Fads abound regarding ways to secure proper ventilation of one's environment. The most suitable "rules" also are the simplest: moderate coolness to remove some body heat without chilling, avoidance of drafts and moderate air circulation to dispel pollutants.

II. TOPICS FOR REVIEW

1. inhalation and exhalation
2. external respiration
3. internal respiration
4. respiratory tract
5. thoracic cavity
6. respiratory rates; variations from normal
7. control of breathing

III. MATCHING EXERCISES

Matching only within each group, print the answer in the space provided.

Group A

cellular (or internal) respiration	larynx	trachea
nasal septum	pharynx	oxygen
external respiration	conchae	carbon dioxide

1. The word "respiration" means "to breathe again"; one of its basic purposes is to supply the body cells with the gas _____.

2. At the same time that the required gas is being supplied, another gas, a waste product of cell metabolism, is being removed. This waste gas is............................ _____.

3. The aspect of respiration involving gas exchanges in the lungs is called .... _____.

4. The second aspect of respiration refers to gas exchanges within the body cells. This is called.................... _____.

5. Below the nasal cavities is a part which is common to both the digestive and respiratory systems. This is the... _____.

6. The cartilaginous structure commonly referred to as the voice box has the scientific name of.................... _____.

7. Several parts of the respiratory tract are kept open by a framework of cartilage. One of these is the windpipe or.. _____.

8. The partition separating the 2 nasal cavities is called the _____.

9. The surface over which the air moves is increased by 3 projections located at the lateral walls of each nasal cavity. These are the .... _____.

Group B

trachea	vascular membrane	nasolacrimal duct
hilum (or hilus)	epiglottis	vocal folds
bronchi	esophagus	sinuses
nasopharynx	oropharynx	

1. The lining of the nasal cavities contains many blood vessels and is therefore described as a.................. _____.

2. The small cavities in the bones of the skull are lined with mucous membrane. They are called.................... _____.

3. Tears are carried from the lacrimal glands across the eye surface, into openings at the corner of the eye, and finally into the nasal cavities by means of a tube called the...... _____.

4. Immediately behind the nasal cavity is the upper portion of the muscular pharynx, the......................... _____.

5. The portion of the pharynx located behind the mouth is the .... _____.

6. The lowest part of the pharynx, the laryngeal pharynx, opens into the air passageway of the larynx, located toward the front, and into the food path, toward the back, where it enters the .... ———————————.

7. The production of speech is aided by the flow of air from the lung to vibrate the.............................. ———————————.

8. Food is prevented from entering the remainder of the respiratory tract by closure of the glottis during swallowing. This is accomplished by a leaf-shaped structure called the .... ———————————.

9. To conduct air to and from the lungs is the purpose of the windpipe, or .... ———————————.

10. The 2 main air tubes to the lungs, formed by division of the trachea, are the............................... ———————————.

11. Each bronchus plus the blood vessels and nerves that accompany it enter the lung at a notch or depression called the .... ———————————.

Group C

diaphragm   squamous epithelium   inhalation
glottis   bronchial tree   exhalation
alveoli   mediastinum   bronchiole

1. Separating the 2 vocal cords is the..................... ———————————.

2. Each bronchus subdivides into many branches. This resemblance to a tree accounts for the name,............. ———————————.

3. The smallest division of a bronchus is called a.......... ———————————.

4. At the end of each terminal bronchiole is a cluster of air sacs, called .... ———————————.

5. Easy passage for the gases that enter and leave the blood in the alveolar capillaries is provided by the very thin air sac wall. This one-cell layer is made of.................. ———————————.

6. The heart is situated in the space between the lungs called the .... ———————————.

7. The physiology of respiration involves 2 phases of breathing. In the first phase air is drawn into the lungs. This is ———————————.

8. In the second phase of breathing air is expelled from the alveoli. This phase is called.............................. ———————————.

9. Separating the thoracic cavity from the abdominal cavity
   is the muscular .... _____ .

IV. LABELING

For each of the following illustrations, print the name or names of each labeled **part**
on the numbered lines.

1. _____

2. _____

3. _____

4. _____

5. _____

6. _____

7. _____

8. _____

9. _____

10. _____

11. _____

12. _____

13. _____

14. _____

15. _____

16. _____

17. _____

18. _____

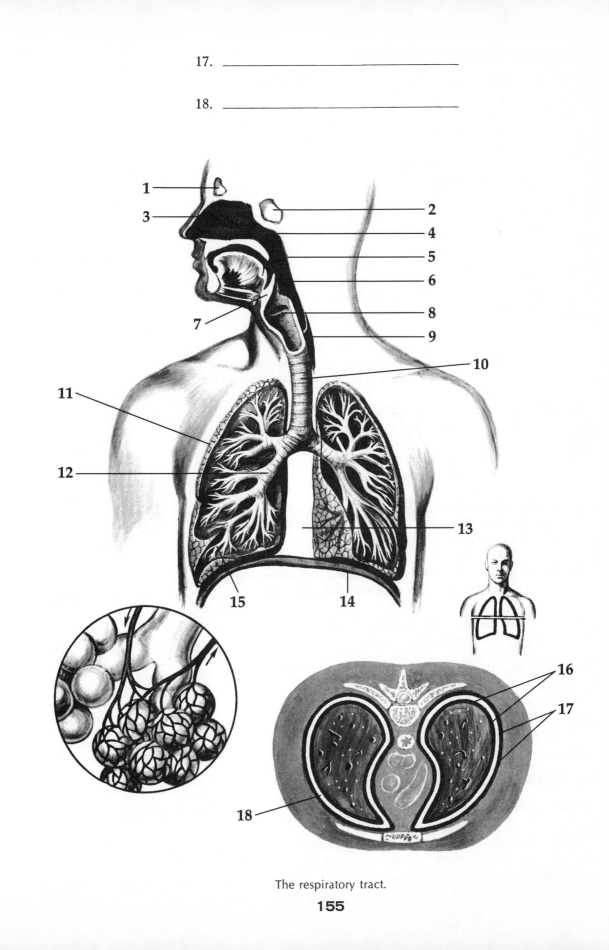

The respiratory tract.

155

V. COMPLETION EXERCISE

Print the word or phrase that correctly completes the sentence.

1. Internal respiration occurs in the microscopic units of body tissues and consequently it is also called.......... _____.

2. The aspect of respiration that involves the inhalation and exhalation of air is................................. _____.

3. Air from the nasal cavities is carried into a space common to the digestive and respiratory systems. This tube is called the .... _____.

4. As a part of the air conditioning accomplished by the nasal mucosa 2 changes made in the inspired air include the addition of .... _____.

5. Dust and other foreign particles are removed from the air by hairs in the nose and by microscopic extensions of the cells of the entire respiratory tract lining. These are called _____.

6. Small pieces of cartilage serve to keep the air spaces open. Several form the framework of the voice box, or......... _____.

7. Near the center of the chest, behind the heart, the trachea divides into 2 tubes called........................... _____.

8. At the end of the smallest subdivisions of the bronchial tree are the clusters of air sacs. The scientific name for air sacs is .... _____.

9. The space between the lungs contains the heart and great vessels, among other things. This space is called the...... _____.

10. The respiratory center for controlling breathing is located in the part of the brain called the..................... _____.

11. The nerve that supplies fibers to the diaphragm is called the .... _____.

12. If there is an increase in the carbon dioxide content of the blood the respiratory center is..................... _____.

VI. PRACTICAL APPLICATIONS

Study each discussion. Then print the appropriate word or phrase in the space provided.

1. Mr. C complained of a severe headache and facial pain. The physician diagnosed the problem as an infection of the air spaces within the cranial bones. These are the.... _____.

2. Young master D, age 5, had a profuse discharge from his nose. There was inflammation of the lining of the nasal cavity. This membrane is made of the tissue called....... _____.

3. Miss F, age 24, complained of difficulty in breathing, partly because of small growths called polyps which were forming between the small lateral projections at the side walls of the nasal cavities. These 3 projections are the.... _____.

4. An additional problem for Miss F was a deviated partition between the 2 nasal spaces. This partition is the.... _____.

5. Miss A, age 43, complained of hoarseness that interfered with teaching and giving instructions. She had an inflammation of her voice box and windpipe. Another name for the windpipe is .... _____.

6. Mrs. D had been suffering from a severe cold and much coughing. X-rays revealed congestion even in the very smallest subdivisions of the tubes of the bronchial tree. These tiny tubes are called........................... _____.

7. Another complication of Mrs. D's cold was a stoppage of the large duct (tube) that carries tears from the region of the eye into the nose. This is the.................... _____.

8. Mr. S, age 54, had recovered from an acute attack of pneumonia. X-rays now showed that there was an accumulation of fluid in the right sac that surrounds the lung. This is the .... _____.

9. Following his recovery from pneumonia and pleurisy, Mr. S complained of sharp pains at the times when he took a deep breath. Adhesions were pulling on the membranes that were connected with the lung and with the lining attached to the chest wall. The layer of membrane attached to the wall has the 2 word name.............. _____.

10. A health worker in the hospital was checking the respiratory rates of the various patients. The normal rates of breathing vary so that each minute a patient may breathe from .... _____.

CHAPTER FOURTEEN

The Urinary System

I. OVERVIEW

The urinary system comprises 2 *kidneys*, 2 *ureters*, 1 *urinary bladder* and 1 *urethra*. This system is usually thought of as the body's main *excretory* mechanism; it is, in fact, often called the *excretory* system. The kidney, however, performs 2 other essential functions: it aids in maintaining *water balance* and in regulating *acid-base balance*.

Prolonged or serious diseases of the kidney nearly always have devastating effects on overall body function and health; a person can live without eyesight and without hearing; but life cannot be maintained unless at least 1 kidney is functioning efficiently. For this reason, renal *dialysis* and renal *transplantation* have been developed in recent years. These methods are helping to save the lives of many persons who otherwise would die of kidney failure and uremia.

II. TOPICS FOR REVIEW

1. excretory mechanisms and interrelationships
   a. digestive system
   b. respiratory system
   c. urinary system
   d. integumentary system
2. kidneys
   a. location
   b. structure
   c. functions
      (1) excretion
      (2) water balance
      (3) acid-base balance
   d. kidney replacements
3. ureters
4. urinary bladder
5. urethra
6. urine, normal constituents

III. MATCHING EXERCISES

Matching only within each group, print the answer in the space provided.

Group A

digestive system   kidneys   elimination
respiratory system   adipose capsule excretion
retroperitoneal space urine

1. Removal of waste products from the body is called...... _____.

2. By contrast, the actual emptying of the hollow organs in which waste substances have been stored is referred to as _____.

3. Other systems besides the urinary system perform excretory functions. To mention one example, bile is excreted by the .... _____.

4. The system regulating excretion of carbon dioxide and water is the .... _____.

5. The urinary system excretes water, nitrogenous waste products and salts, all of which are contained in the..... _____.

6. Extraction of wastes from the blood is a function of the.. _____.

7. The area behind the peritoneum which contains the pancreas, duodenum and the 2 kidneys is referred to as the.. _____.

8. The circle of fat that helps to support the kidney is called the .... _____.

Group B

collecting tubules epithelium   filtration
renal basin   Bowman's capsule urea
cortex   convoluted tubule reabsorption
glomerulus

1. The cluster of capillaries located at one end of the nephron is the .... _____.

2. Materials that have passed through the capillary walls enter the first part of the nephron, the.................... _____.

3. The nephron is basically a tiny coiled tube called a...... _____.

4. The useful substances that have escaped through the nephron capillaries are sent back to the bloodstream by a process of .... _____.

5. Since the kidney is a gland, it is made up mainly of...... _____.

6. The combination of glomerulus and Bowman's capsule connected with each of the one million nephrons of the kidney provides a highly effective means of............ _____.

7. The part of the kidney containing the nephron bulbs and their blood vessels is the............................. _____.

8. Within the medulla the open ends of the nephron tubes come together (and empty into) the.................... _____.

9. The urine produced by structures located in the cortex is collected by tubules in the medulla. These latter tubules empty into the pelvis, or........................... _____.

10. As body cells use protein, nitrogenous waste products are produced; the chief such product is.................... _____.

Group C

mineral salts	internal sphincter	peristalsis
glucose	urethra	hilum
calyces	organic	acid-base balance

1. Urine is moved along the ureter from the kidneys to the bladder by the rhythmic contraction known as .... _____.

2. Near the bladder outlet are circular muscle fibers that contract to prevent emptying. They form what is known as the .... _____.

3. The tube that carries urine from the bladder to the outside is the .... _____.

4. Because nitrogen waste products originate from living organisms they are said to be .... _____.

5. Inorganic compounds normally contained in urine are also classified as .... _____.

6. Diabetes mellitus may be suspected if a test of the urine shows the presence of the simple sugar .... _____.

7. The area where the artery, the vein and the ureter connect with the kidney is known as the .... _____.

8. Tubelike extensions that project from the renal pelvis into the kidney tissue serve to increase the area for collection of urine. These extensions are called .... _____.

9. The kidney helps prevent conditions of excessive alkalinity or acidity by regulating the body's .... _____.

Group D

ureters	acid	retroperitoneal space
buffers	salts	membranous capsule
nephrons	urea	water balance
kidneys	urethra	peristalsis

1. The body's intake and output of liquid must be carefully regulated at all times to maintain a normal state of health. The kidney is an organ that aids in maintaining . . . . _____.

2. The kidneys are actually located outside the peritoneal cavity in an area called the . . . . _____.

3. Each kidney is enclosed in a loosely adherent fibrous connective tissue structure called the . . . . _____.

4. All types of kidney inflammation are largely characterized by destruction of the functional units which are microscopic and are called . . . . _____.

5. The chief nitrogen waste product is formed mostly from the metabolism of food proteins. It is named . . . . _____.

6. Unless replaced by certain machines, the organs of the urinary system that are absolutely necessary for life are the glandular . . . . _____.

7. The excretory tube of the bladder is the . . . . _____.

8. Urine is normally conducted from the kidneys to the bladder by tubes called . . . . _____.

9. The rhythmic contraction of muscles in the walls of the ureters is called . . . . _____.

10. Exhalation of carbon dioxide is one means by which there is removal of substances that are . . . . _____.

11. Acids are neutralized by alkalis to form . . . . _____.

12. The nearly neutral state of the blood is due to the presence of the mineral salts known as . . . . _____.

IV. LABELING

For each of the following illustrations, print the name or names of each labeled part on the numbered lines.

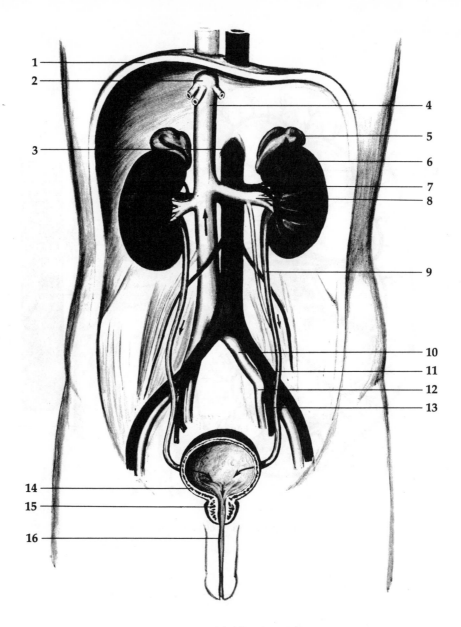

Urinary system with blood vessels.

1. _____ 7. _____ 13. _____

2. _____ 8. _____ 14. _____

3. _____ 9. _____ 15. _____

4. _____ 10. _____ 16. _____

5. _____ 11. _____

6. _____ 12. _____

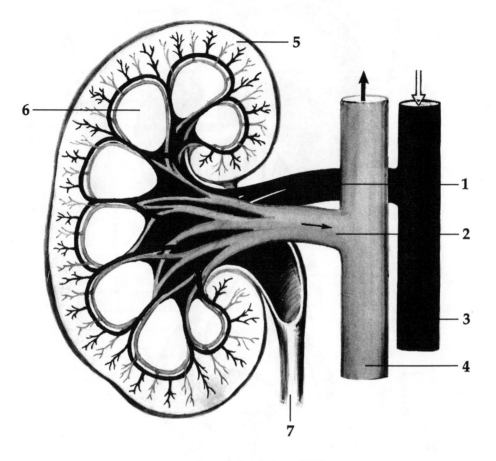

Blood supply and circulation of kidney.

1. _____

2. _____

3. _____

4. _____

5. _____

6. _____

7. _____

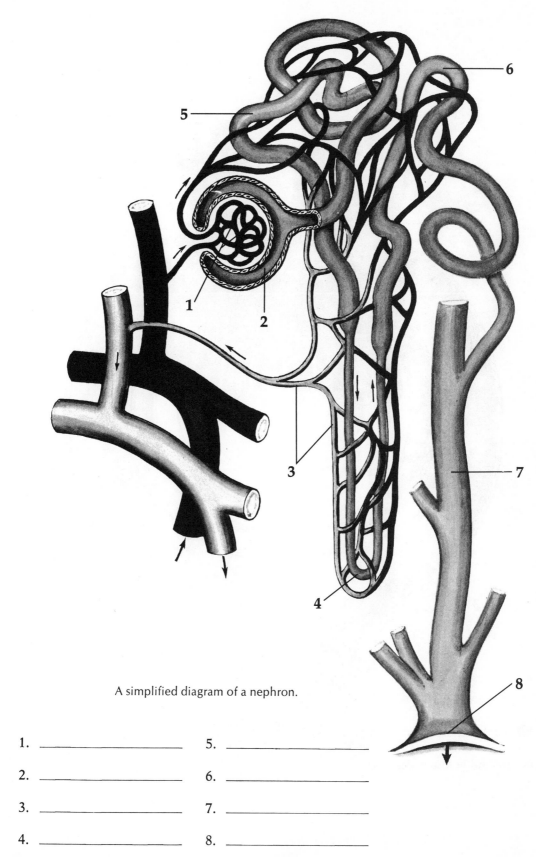

A simplified diagram of a nephron.

1. _____   5. _____

2. _____   6. _____

3. _____   7. _____

4. _____   8. _____

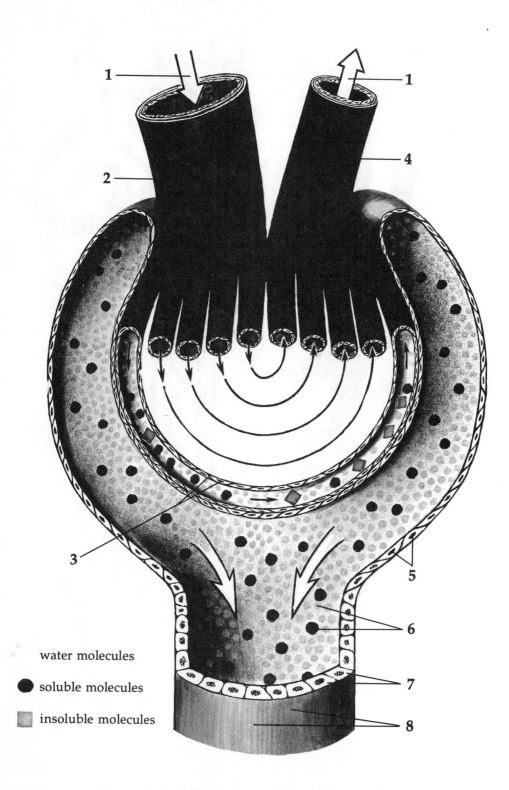

water molecules

● soluble molecules

▦ insoluble molecules

Diagram to show filtration process during formation of urine.

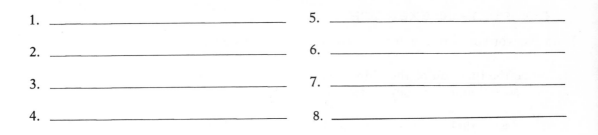

1. _____   5. _____

2. _____   6. _____

3. _____   7. _____

4. _____   8. _____

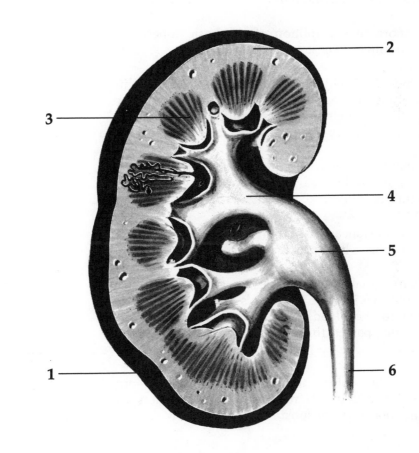

Kidney's internal structure.

1. _____   4. _____

2. _____   5. _____

3. _____   6. _____

V. COMPLETION EXERCISE

Print the word or phrase that correctly completes the sentence.

1. Since the function of the kidney includes the production of a secretion, this organ is often described as .... _____.

2. The vessel that carries oxygenated blood to the kidney is the .... _____.

3. A category of chemical substances that neutralize acids are the bases, also called .... _____.

4. Regulating the acid-base balance of the body is an important function of the 2 glandular .... _____.

5. There are more than a million of the basic functional units in each kidney. These units are called .... _____.

6. In order that blood may be acted upon by the kidney in as large amounts as possible, each nephron has a cluster of capillaries at its beginning. This capillary tuft is known as the .... _____.

7. At the beginning of each convoluted tubule of the nephron is a cuplike portion that nearly surrounds the glomerulus. This structure is called .... _____.

8. A number of substances pass from the blood into the first part of the nephron by a process called .... _____.

9. Since certain valuable substances filter through the glomerular capillaries into the Bowman's capsule, some must be returned to the blood. This is accomplished as the filtered mixture passes through the much coiled tubule by a process of .... _____.

10. At the end of the nephron the mixture has become more concentrated and contains nitrogen waste products as well as other nonessential substances. This is now .... _____.

VI. PRACTICAL APPLICATIONS

Study each discussion. Then print the appropriate word or phrase in the blank space provided.

1. An x-ray examination of Mrs. L's kidneys revealed destruction of the tubelike extensions from the kidney pelvis. These projections from the pelvis into the medulla are known as .... _____.

2. Congenital abnormalities of the urinary tract often involve the tubes that carry urine from the kidneys into the urinary bladder. These are the .... ———————————.

3. Mr. R was suffering from nephritis, or inflammation of the kidney. In this disorder there is destruction of the very important functional units called .... ———————————.

4. A test of Mr. R's blood revealed an abnormally high content of the chief nitrogen waste product called .... ———————————.

5. Mrs. K complained of painful urination and examination revealed an inflammation of the bladder called cystitis. Normally the empty bladder's lining is thrown into folds called .... ———————————.

6. Mr. J, age 58, complained of difficulty in emptying his bladder. Examination (with a cystoscope) showed an enlargement of the prostate gland. The male sex cells are carried into the first part of the excretory tube called the ———————————.

7. Investigation of Mr. J's urinary problem included the passage of tubes (catheters) up through the urethra and urinary bladder, and finally through the tubes, the ureters, into the kidney basin, also called the .... ———————————.

8. The male urethra is much longer than the female. In centimeters it measures about .... ———————————.

9. Mr. J had problems in emptying his bladder. This process is called urination or .... ———————————.

10. The control of urination involves training the reflex present in the infant so that the circular muscle fibers contract to prevent emptying of the bladder. These muscle fibers form a valvelike structure called the .... ———————————.

11. Mrs. K suffered from an inflammation of the urethral lining as well as of the bladder. The female urethra serves but one purpose, that of conducting the excretory liquid from the bladder. This liquid is .... ———————————.

12. The excretion from the bladder was examined in the case of each of these patients. The main constituent, about 95%, is .... ———————————.

CHAPTER FIFTEEN

Glands and Hormones

I. OVERVIEW

Glands are organs that manufacture secretions. The glands are divided into 2 types: the *exocrine* glands have ducts that carry the secretion to other parts of the body; the *endocrine* glands are ductless, and the blood and the lymph carry their secretions. The secretions are also divided into 2 classes: the *external* secretions are carried from the gland cells to a nearby organ or to the body surface; the *internal* secretions are carried to all parts of the body by the blood or the lymph.

The endocrine glands manufacture *hormones*, chemical substances that regulate the activities of various body organs. These hormones perform many essential functions, of which control of *body growth*, control of *food metabolism* and control of the growth and development of the *sexual organs* are examples. So important, in fact, is the work of the hormones that they are often referred to as the body's *chemical messengers*.

Among the hormones that are being studied intensively at the present time are the neurohormones and the prostaglandins. Much remains to be learned about these and other hormones, the tissues that produce them and the effects they cause within the body.

II. TOPICS FOR REVIEW

1. secretions
   a. external
   b. internal
2. glands
   a. exocrine
   b. endocrine
3. hormones
   a. thyroid gland; thyroxine
   b. parathyroid gland; parathormone
   c. pituitary gland; anterior lobe, posterior lobe
   d. pancreas
   e. adrenal glands
   f. sex glands

4. placenta
5. thymus
6. pineal body
7. neurohormones and prostaglandins

III. MATCHING EXERCISES

Matching only within each group, print the answer in the space provided. The same answer may be used more than once.

Group A

iodine	basal metabolism	islands (or islets) of
external secretions	hormones	Langerhans
parathyroid glands	thyroid	suprarenal glands

1. The digestive juices and tears are examples of .... _____.

2. The substances produced by endocrine glands are known as .... _____.

3. Glands that have ducts produce substances classified as .. _____.

4. The body's "chemical messengers" are the .... _____.

5. The adrenal glands are also known as the .... _____.

6. The groups of specialized cells scattered throughout the pancreas are known as the .... _____.

7. The largest of the endocrine glands is located in the neck. It is the .... _____.

8. One test for evaluating thyroid function is done when the person is at complete rest. This test determines the person's .... _____.

9. Located behind the thyroid gland and embedded in its capsule are the 4 .... _____.

10. A relatively simple test for thyroid function involves taking blood from a vein and then testing it for the so-called protein-bound .... _____.

Group B

pituitary	medulla	adrenal
goiter	placenta	insulin
thyroxine	iodine	parathormone

1. Production of heat and energy in the body tissues is regulated by the hormone .... ————————.

2. In order that thyroxine may be manufactured, the blood must contain an adequate supply of .... ————————.

3. The adrenal glands as well as the kidneys and other organs have an inner part called the .... ————————.

4. The amount of calcium dissolved in the circulating blood is regulated by the parathyroid secretion .... ————————.

5. Several essential hormones are produced by the anterior and posterior lobes of the .... ————————.

6. In order to provide for normal sugar utilization in the tissues the islands of Langerhans must produce the hormone .... ————————.

7. The external cortex and the internal medulla act as separate glands with specific functions in the case of the ... ————————.

8. The normal development of the embryo is aided by hormones from the ovaries, pituitary, and an organ present only during pregnancy, namely the .... ————————.

9. In the disorder known as diabetes mellitus sugar is not "burned" in the tissues to produce energy. This is due to a lack of the hormone .... ————————.

Group C

aldosterone	oxytocin	ACTH
endocrine	estrone	progesterone
cortisol	exocrine	epinephrine

1. During stressful situations such as injury or surgery the the body is protected somewhat by a hormone (a glucocorticoid) that is usually called .... ————————.

2. Contraction of the pregnant uterus is stimulated by a hormone from the posterior pituitary called .... ————————.

3. Blood pressure is raised and the rate of the heartbeat is increased by the chief hormone of the adrenal medulla ... ————————.

4. The reabsorption of sodium in the kidney tubules is an important electrolyte-regulating function of the adrenal cortex hormone .... ————————.

173

5. Testosterone is produced by the male sex glands; the female sex glands produce a hormone which most nearly parallels testosterone in its action. This hormone is called _____.

6. The lacrimal gland is an example of one that is described as .... _____.

7. A hormone that is necessary for normal development of pregnancy is one produced by the female sex glands. It is called .... _____.

8. When the needs of the body are such that amino acids must be changed to sugar instead of protein, the adrenal cortex produces large amounts of the hormone .... _____.

9. The pituitary is stimulated by impulses from the hypothalamus, while the adrenal cortex is stimulated by the anterior pituitary hormone known as.................. _____.

10. The somatotropic hormone is an example of one of many hormones that are manufactured by glands that are classified as .... _____.

IV. LABELING

Print the name or names of each labeled part on the numbered lines.

1. _____

2. _____

3. _____

4. _____

5. _____

6. _____

7. _____

8. _____

9. _____

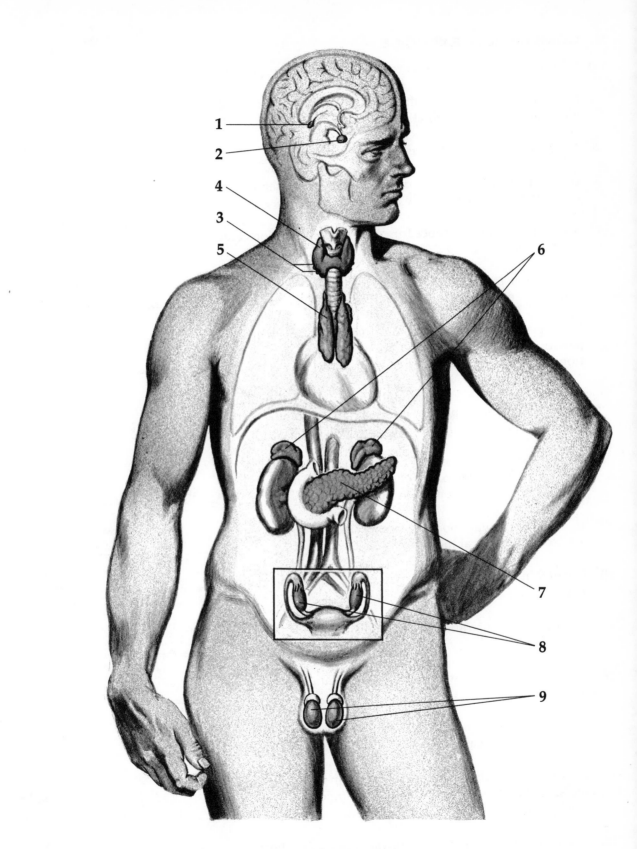

Endocrine system.

V. COMPLETION EXERCISE

Print the word or phrase that correctly completes the sentence.

1. Substances that are manufactured by the special cells of glandular tissue are called .... _____.

2. An external secretion that is also an excretion is .... _____.

3. Glands that have tubes or ducts to carry their secretions to another part of the body are .... _____.

4. Some glands have no tubes for carrying the secretions and so are called .... _____.

5. In addition to the framework of connective tissue most glands are made of .... _____.

6. Many glands have fibrous connective tissue capsules with partitions extending into the organ. Between these partitions are units of gland tissue called .... _____.

7. The mucosa of the stomach and small intestine produce external secretions plus a hormone that stimulates the pancreas and other glands. This hormone is .... _____.

8. Among the most vascular of organs are the 2 small glands located at the upper poles of the kidneys. These are the adrenal or more appropriately the .... _____.

9. In order to regulate body processes the nervous system is assisted by chemical messengers called .... _____.

10. Located in the neck is the largest of the endocrine glands. It is the .... _____.

11. The rate of glucose absorption and utilization is increased by the hormone .... _____.

12. The blood iodine test is used to determine the activity of the .... _____.

13. Embedded in the capsule of the thyroid gland are 4 small epithelial bodies called the .... _____.

14. Located just behind the optic chiasma in a depression of the sphenoid bone is a small but very important so-called master gland named the .... _____.

15. The anterior lobe of the hypophysis produces a hormone that stimulates the production of milk. It is the .... _____.

VI. PRACTICAL APPLICATIONS

Study each discussion. Then print the appropriate word or phrase in the space provided.

1. Young master J, age 10, was being studied for a possible abnormality of the pituitary gland. He was very small for his age and the possibility that he would be a midget was suggested. Another name for the growth-promoting hormone is .... ——————————.

2. The growth-promoting hormone is produced by the part of the pituitary called the .... ——————————.

3. Following the birth of her child Mrs. M was given a hormone that would stimulate the contraction of the muscles of the uterus. The extract used for this purpose is called ——————————.

4. The extract used for Mrs. M to stimulate the uterine musculature comes from the part of the pituitary called the .. ——————————.

5. Mrs. C was brought to the hospital in a coma, that is, she was unconscious and could not be aroused. Tests revealed that she had a very high blood sugar. Mrs. C's illness was due to a lack of insulin, a hormone produced by parts of the pancreas called the .... ——————————.

6. The 2-word name for the disorder that Mrs. C was suffering from is .... ——————————.

7. Mr. F was suffering from a disorder of the kidneys so severe that it was necessary to use the artificial kidney, a dialysis machine. There was insufficient absorption of potassium. The principal hormone responsible for electrolyte regulation under normal circumstances is produced by the adrenal cortex and is called .... ——————————.

8. In the case of Mrs. D, it was learned that she had had tuberculosis during her "teens." Now she was unable to withstand stressful situations and her resistance to infection was low. It was believed that there was damage to the outer part of the adrenal gland. This portion of the gland is the .... ——————————.

9. Seventeen-year-old Miss K had not had a menstrual period. A possible cause was thought to be a lack of the ovarian hormone called .... ——————————.

10. Baby Y had been extremely susceptible to all kinds of infections. A structure that is most active during prenatal life and in infancy, and one that is involved in the production of lymphocytes, is located in the upper chest and is called the .... ——————————.

CHAPTER SIXTEEN

The Reproductive System

I. OVERVIEW

Through the functioning of the reproductive system, the continuation of the human race is assured. Human reproduction is *sexual* (as compared to the *asexual* reproduction of some of the simplest forms of life).

The male reproductive system consists of the *testes*, the *seminal vesicles*, the *prostate gland*, the *penis* and the *bulbourethral glands*; in the female this system comprises the *ovaries*, the *fallopian tubes*, the *uterus*, the *vagina*, the *vulva*, the *greater vestibular glands* and the *perineum*.

Normally, *fertilization* of a female *ovum* by a male *spermatozoon* results in *pregnancy*, the period of about 9 months during which an *embryo* forms and develops into a *fetus*. The ability of a woman to bear children ends with the menopause, the final cessation of menstruation.

II. TOPICS FOR REVIEW

1. sexual and asexual reproduction
2. characteristics common to male and female reproductive systems
3. male reproductive system
   a. testes
   b. tubes
   c. seminal vesicles
   d. prostate gland
   e. urethra and penis
   f. mucous glands
   g. spermatozoa
4. female reproductive system
   a. ovaries
   b. fallopian tubes
   c. uterus
   d. vagina
   e. vulvovaginal glands
   f. vulva and perineum

5. pregnancy
   a. first stages
   b. the embryo
   c. the fetus
   d. parturition
   e. mammary glands; lactation
   f. multiple births
6. menopause

III. MATCHING EXERCISES

Matching only within each group, print the answer in the space provided. An answer may be used more than once.

Group A

ovum	ovary	epididymis
gonads	asexual	testosterone
testis	scrotum	sexual
spermatozoa		

1. The specialized sex cells in the male are called .... _____

2. Since the simplest forms of life require no partner in order to reproduce, they are said to be .... _____

3. A characteristic shared by both men and women is the presence of sex glands, or .... _____

4. The specialized sex cell of the female is the .... _____

5. The female gonad is also known as the .... _____

6. The male gonad is the .... _____

7. The testes are normally located in a sac that is suspended between the thighs. This sac is the .... _____

8. The bulk of the tissue of the testes is arranged in tubules. Within the walls of these tubules there is the production of .... _____

9. Groups of cells located between the tubules of the testes are responsible for the secretion of the male hormone ... _____

10. The spermatozoa mature and become motile within a temporary storage area, a 20-foot tube, the .... _____

11. In most animals, there is differentiation into male and female; reproduction is therefore said to be .... _____

Group B

seminal vesicles   ductus deferens   sex glands
ejaculatory ducts   gametes   urethra
spermatic cord   penis

1. The gonads are the .... ————————.

2. Another term used in referring to the male and female
   sex cells is .... ————————.

3. The straighter upward extension of the epididymis is the
   beginning of the .... ————————.

4. The combination of ductus deferens, the nerves and the
   blood and lymph vessels that extend from the scrotum
   and testes on each side is named the .... ————————.

5. Behind the urinary bladder in the male are 2 tortuous
   muscular tubes with glandular linings. These are the .... ————————.

6. The vas deferens may also be called the .... ————————.

7. The spermatozoa are carried in the seminal vesicle secretion through the prostate gland via 2 tubes called the .... ————————.

8. A single tube conveys urine from the bladder and carries
   reproductive cells to the outside. This tube is the .... ————————.

9. The external genitalia of the male include the scrotum
   and the .... ————————.

10. Spermatozoa are nourished by secretions produced by
    the glandular lining of the .... ————————.

Group C

gametes   scrotum   abdominal wall
penis   semen   vas deferens
prepuce   erection   Cowper's glands
testes   pelvic cavity

1. In male ejaculation, a mixture of spermatozoa and secretions is expelled which is called .... ————————.

2. In the male, the longest part of the urethra extends
   through the spongelike .... ————————.

3. The bulbourethral glands, which are pea-sized organs
   found in the pelvic floor tissues below the prostate gland
   of the male, are also known as .... ————————.

4. Spermatozoa are produced by mitosis and meiosis within the tubules that form much of the .... _____.

5. The 2 spermatic cords extend from the testes in the scrotum up on each side and then through the .... _____.

6. In the embryo the gonads descend from the region of early development near the kidneys. At birth both the male and female gonads will have normally completed the descent. The ovaries descend only as far as the .... _____.

7. The descent of the male gonads is much farther than in the female, since the testes normally come to be finally located outside the body proper in a special sac called the _____.

8. The foreskin is also called the .... _____.

9. Another name for sex cells or germ cells is .... _____.

10. In order to prevent the spermatozoa from reaching the urethra, purposeful sterilization of the male is accomplished by removing a portion of the .... _____.

11. Semen is expelled into the female vagina through the stiffening of the penis known as .... _____.

Group D

vulva	vagina	fallopian tubes
ovaries	ovarian follicles	uterus
ovulation	vulvovaginal glands	fimbriae

1. Two structures, made of peritoneum and called the broad ligaments, serve as anchors for the .... _____.

2. The sacs within which the ova mature are called the .... _____.

3. The rupture of an ovarian follicle permits an ovum to be discharged from the ovary surface. This is called .... _____.

4. The mature ovum travels from the region of the ovary into the oviducts or .... _____.

5. A current in the peritoneal fluid sweeps the ovum into the oviduct. This current is produced by the fringelike ...... _____.

6. Before birth the fetus grows in a muscular organ located between the urinary bladder and the rectum. This organ is the .... _____.

7. Bartholin's glands, situated above and to each side of the vaginal opening, are also known as the greater vestibular glands, or the .... ————————.

8. Connecting the uterus with the outside is the lower part of the birth canal, the .... ————————.

9. The labia, the clitoris and related structures comprise the external parts of the female reproductive system which are called the .... ————————.

Group E

broad ligaments	anus	uterus
endometrium	cervix	perineum
pituitary (hypophysis)	fundus	corpus
		fornices

1. Located above the level of the tubal entrances is the small rounded part of the uterus called the .... ————————.

2. The upper part of the uterus is the larger part; it is called the body, or .... ————————.

3. The necklike part of the uterus dips into the upper vagina; this necklike part is called the .... ————————.

4. The specialized tissue that lines the uterus is known as .. ————————.

5. The cervix dips into the upper vagina so that a circular recess is formed; this gives rise to the spaces known as ... ————————.

6. The pelvic floor in both the male and the female is properly called the .... ————————.

7. The structures that are made of 2 layers of peritoneum located at each side of the uterus are known as the ..... ————————.

8. The production of gametes in both sexes is apparently influenced by hormones from the anterior lobe of the master gland, the .... ————————.

9. During the process of birth the part of the perineum that could be damaged is that region between the vaginal opening and the .... ————————.

10. The oviducts (fallopian tubes) are not connected with the ovaries but only with the upper angles of the .... ————————.

Group F

parturition	umbilicus	embryo
fetus	placenta	labor pains
afterbirth	vernix caseosa	corpus luteum
amniotic sac		

1. The endometrium is prepared for the fertilized ovum by the hormone progesterone, which is produced by the .... _____.

2. Serving as the organ for nutrition, respiration and excretion for the embryo is a flat, circular structure called the _____.

3. Following fertilization of an ovum and until the end of the third month, the developing organism is called the ... _____.

4. From the end of the third month until birth the developing organism is known as the .... _____.

5. The fetus is protected by a fluid produced by the membrane lining the .... _____.

6. Nature provides various protective mechanisms for the fetus. The cheesy material that protects the skin is known as .... _____.

7. The process of giving birth to a child is described by the term .... _____.

8. A small part of the umbilical cord remains attached to the navel for a few days following birth. The scientific name for the navel is the .... _____.

9. Normally, within half an hour after the child is born, the placenta together with the membranes of the amniotic sac and most of the umbilical cord are expelled as the ...... _____.

10. Parturition is usually divided into 3 stages. In the first stage the muscles of the uterus begin the contractions known as .... _____.

Group G

menstrual flow	ovulation	uterus
estrone	posterior fornix	vagina
menopause	external genitalia	cilia

1. The rupture of an ovarian follicle followed by the escape of an ovum is called .... _____.

184

2. The movement of an ovum from the abdominal end of the oviduct toward the uterine cavity is made possible by the waving of the microscopic extensions of the lining cells. These are the .... ——————————————.

3. Both ovarian hormones are involved in preparing the endometrium for pregnancy and both are carried by the bloodstream to the uterus. The preparation is initiated by the hormone found in the fluid surrounding the maturing ovum. This hormone is .... ——————————————.

4. The peritoneal cavity of the female is deepest behind the upper vaginal canal. This means that there is a thin wall separating the lower abdominal cavity from the upper vaginal canal. This dorsal space in the upper vagina is called the .... ——————————————.

5. Bits of cast-off endometrium are found in the bloody discharge that is known as the .... ——————————————.

6. The vulva is also called the .... ——————————————.

7. The muscular pear-shaped organ located within the female pelvis is the .... ——————————————.

8. Cessation of ovarian activity brings about the period of life known as .... ——————————————.

9. Folds in the mucosal lining are characteristic of the stomach, the urinary bladder and the .... ——————————————.

IV. LABELING

For each of the following illustrations, print the name or names of each labeled part on the numbered lines.

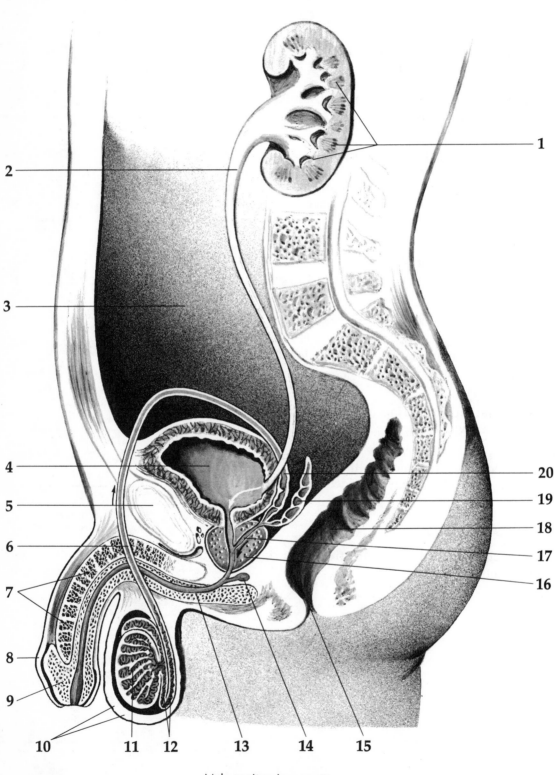

Male genitourinary system.

186

1. _____

2. _____

3. _____

4. _____

5. _____

6. _____

7. _____

8. _____

9. _____

10. _____

11. _____

12. _____

13. _____

14. _____

15. _____

16. _____

17. _____

18. _____

19. _____

20. _____

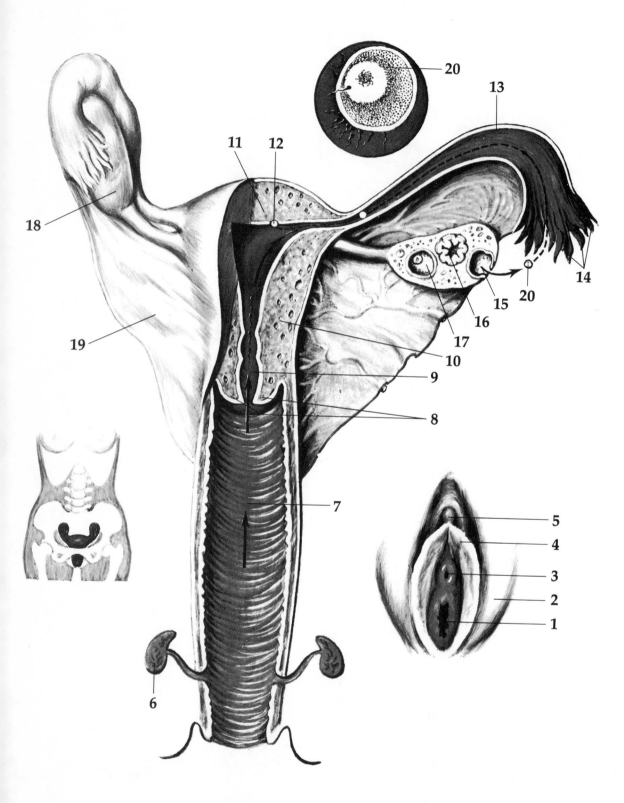

Female reproductive system.

1. _____

2. _____

3. _____

4. _____

5. _____

6. _____

7. _____

8. _____

9. _____

10. _____

11. _____

12. _____

13. _____

14. _____

15. _____

16. _____

17. _____

18. _____

19. _____

20. _____

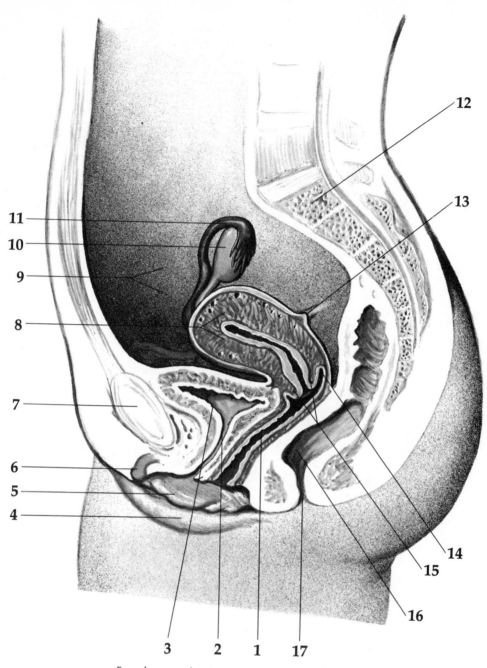

Female reproductive system, sagittal section.

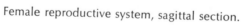

1. _____

2. _____

3. _____

4. _____

5. _____

6. _____

7. _____

8. _____

9. _____

10. _____

11. _____

12. _____

13. _____

14. _____

15. _____

16. _____

17. _____

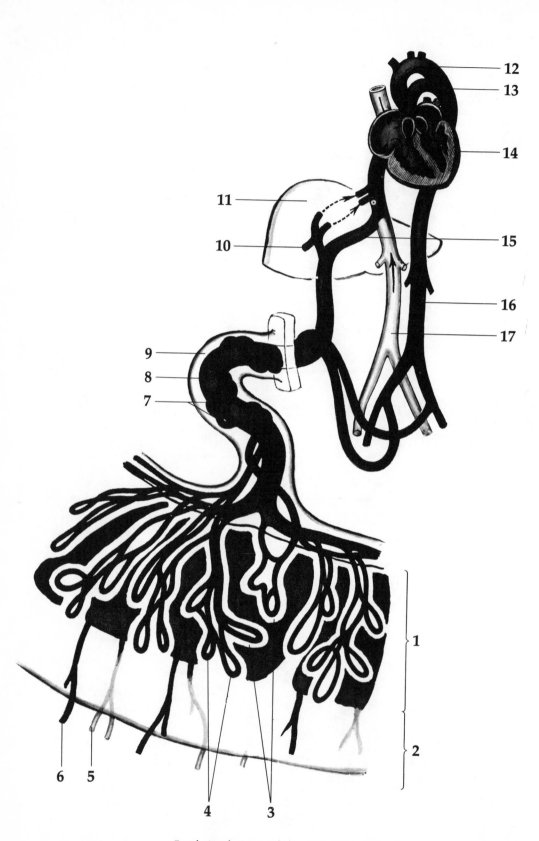

Fetal circulation and the placenta.

1. _____

2. _____

3. _____

4. _____

5. _____

6. _____

7. _____

8. _____

9. _____

10. _____

11. _____

12. _____

13. _____

14. _____

15. _____

16. _____

17. _____

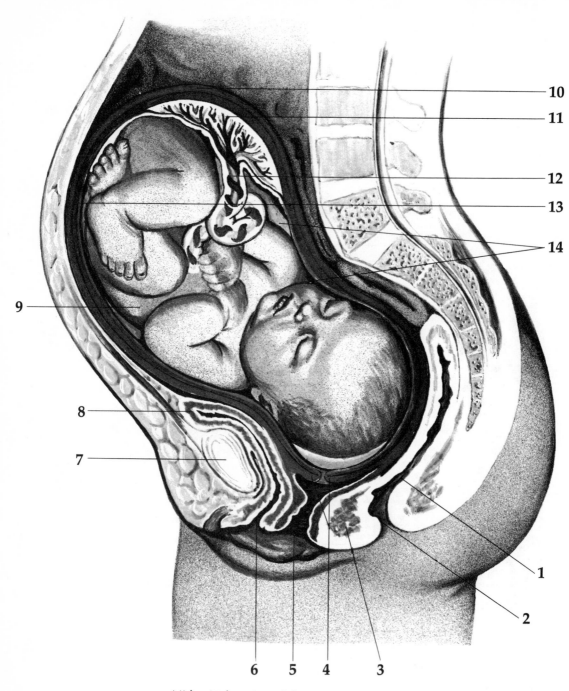

Midsagittal section of the pregnant uterus.

1. _____   4. _____

2. _____   5. _____

3. _____   6. _____

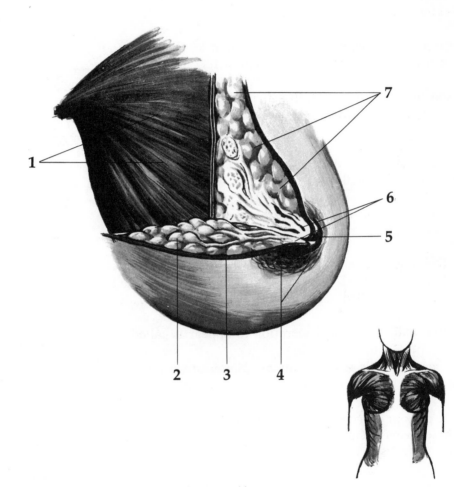

Section of breast.

1. _____

2. _____

3. _____

4. _____

5. _____

6. _____

7. _____

V. COMPLETION EXERCISE

Print the word or phrase that correctly completes the sentence.

1. The motility of the spermatozoa is maintained by several secretions including one from the seminal vesicles and one produced by the prostate gland called the .... _____.

2. Rupture (hernia) is much more common in the male. It occurs mainly in the inguinal region where a combination of tubes and vessels extend through the abdominal wall. This combination is called the........................ _____.

3. The individual spermatozoon is very motile. It is able to move toward the ovum by the action of its .... _____.

4. By the end of the first month of embryonic life the beginnings of the extremities may be seen. These are 4 small swellings called .... _____.

5. The cell formed by the union of a male sex cell and a female sex cell is called a................................. _____.

6. The science that deals with the development of the embryo is called .... _____.

7. The bag of waters is a popular name for the membranous sac that encloses the fetus. The clear liquid that is released during labor is called .... _____.

8. The first mammary gland secretion to appear is called... _____.

9. The secretory cells of the mammary glands are stimulated by a pituitary hormone called .... _____.

10. About once in every 80 to 90 births twins are born. Some of these twins occur as a result of 2 different ova being fertilized by 2 spermatozoa. Such twins are said to be .... _____.

11. Some twins develop from a single zygote formed from a single ovum that has been fertilized by a single spermatozoon. The embryonic cells separate during early stages of development. These twins are described as .... _____.

12. The mammary glands of the female provide nourishment for the newborn through the secretion of milk; this is called .... _____.

13. Normally uterine contractions are inhibited during pregnancy by a secretion from the placenta called .... _____.

14. The beginning of the uterine contractions of childbirth may be due to the hormones produced by the placenta, called .... ——————————————.

VI. PRACTICAL APPLICATIONS

Study each discussion. Then print the appropriate word or phrase in the space provided.

1. Mr. K, age 18, was complaining of discharge of pus from the urethra and of pain in the scrotum. The physician found that he had an infection of the long (6 meters) much coiled tube that carries spermatozoa from the testis. This is the .... ——————————————.

2. Mr. J, age 58, came in to his physician complaining of difficulty in urinating. Examination revealed an enlargement of a gland that is often affected in this way in men past middle age. This is the .... ——————————————.

3. Mr. J had a temperature of 102°F (39°C). In addition to the enlargement of the gland surrounding the first part of the urethra there was also infection that involved the tortuous muscular tubes and their outpouchings. These are the .... ——————————————.

4. Couple K came in for evaluation of a seeming sterility. After 12 years of marriage they had no children. It is recommended in cases of this kind that the male be checked first since this is easier and at any rate must be done for a complete study. The first check was of the detached cells in the seminal fluid, the .... ——————————————.

5. Mrs. K required a more difficult examination including an evaluation of the oviducts. Another name for these structures is uterine tubes or .... ——————————————.

6. It was found that Mrs. K had a small tumor, a myoma (or fibroid), in the uppermost part of the uterus. This small rounded part located above the level of the tubal entrances is the .... ——————————————.

7. A microscopic examination of the cells of the uterine lining was also included for Mrs. K. This uterine lining is named .... ——————————————.

8. Mrs. Y, age 27, was pregnant a second time. Her first child suffered from a disorder known as Down's syndrome. It was decided that a study of the fluid in the sac surrounding the embryo was in order. This fluid is called .... ——————————————.

197

References

Memmler, R. L., and Wood, D. L.: Structure and Function of the Human Body, ed. 2. Philadelphia, Lippincott, 1977.

Memmler, R. L., and Wood, D. L.: The Human Body in Health and Disease, ed. 4. Philadelphia, Lippincott, 1977.

Anthony, C. P.: Basic Concepts in Anatomy and Physiology, ed. 3. St. Louis, Mosby, 1974.

Brooks, S.: Basic Science and The Human Body: Anatomy and Physiology. St. Louis, Mosby, 1975.

Chaffee, E. E., and Greisheimer, E. M.: Basic Physiology and Anatomy, ed. 3. Philadelphia, Lippincott, 1974.

Crouch, J. E., and McClintic, J. R.: Human Anatomy and Physiology, New York, Wiley, 1971.

Greisheimer, E. M., and Wiedeman, M. P.: Physiology and Anatomy, ed. 9. Philadelphia, Lippincott, 1972.

Griffiths, M.: Introduction to Human Physiology. New York, Macmillan, 1974.

Lenihan, J.: Human Engineering; The Body Re-examined. New York, Braziller, 1975.

Shepro, D.: Human Anatomy and Physiology. New York, Holt, 1974.

Stonehouse, B.: The Way Your Body Works. London, Beasley, 1974.